I0038564

Portfolio and Investment Analysis with SAS®

Financial Modeling Techniques for Optimization

John B. Guerard
Ganlin Xu
Ziwei Wang

sas.com/books

The correct bibliographic citation for this manual is as follows: Guerard, John B., Ziwei Wang, and Ganlin Xu. 2019. *Portfolio and Investment Analysis with SAS®: Financial Modeling Techniques for Optimization.* Cary, NC: SAS Institute Inc.

Portfolio and Investment Analysis with SAS®: Financial Modeling Techniques for Optimization

978-1-64295-193-6 (Hard cover)
978-1-63526-692-4 (Hardcopy)
978-1-63526-691-7 (Web PDF)
978-1-63526-689-4 (epub)
978-1-63526-690-0 (mobi)

SAS Institute Inc., SAS Campus Drive, Cary, NC 27513-2414

March 2019

Contents

About This Book

What Does This Book Cover?

In this applied investment book, we introduce the risk-return tradeoff of financial investments as well as stocks and bonds investing in the US and global markets over the 2002–2016 time period. Stocks have produced higher rates of return relative to risk in global markets than US markets. We report why intelligent investors prefer more stocks than bonds to maximize stockholder wealth. The bulk of this book is concerned with demonstrating how individuals, whether students or real-world investors, can select stocks and create portfolios to maximize expected returns for a given level of risk. The authors do not believe in completely Efficient Markets, and we show how to outperform the markets by using sophisticated statistical modeling implemented in SAS. The authors believe that "Quant" life is pass/fail. Your models are either statistically significant, or they are not.

The authors have used SAS for over 35 years in financial modeling and investment research. We stress the need to generate statistically significant stock selection modeling and portfolio construction management and measurement. The authors believe that the Markowitz Efficient Frontier can be applied by students, investors, and Certified Financial Planners using SAS to create variables, run robust regression models for stock selection, and use the stock expected returns and risk inputs to create Efficient Frontiers.

Is This Book for You?

We assume no previous knowledge of finance, investments, statistics, or optimization. The text shows you how to analyze income statements, balance sheets, and sources and uses of funds statement analyses. We show why some variables are more useful to consider for model building based on Information Coefficients (ICs) and estimated Efficient Frontiers with realistic transactions costs. The authors discuss stock universes developed and modeled from the perspective of ICs and decile spreads. We report why variables might be different in US, Chinese, Japanese, European, Emerging Market, and global stock universes with respect to a large set of variables of analysts' earnings per share forecasts, forecast revisions, and direction of revisions. The authors have written technical papers and texts that are included in the References section. In this book, we want to introduce analysis beyond the typical undergraduate investments course to help enhance portfolio returns.

We use PROC REG, PROC IML, and PROC ROBUSTREG to run monthly regressions and to demonstrate how to address outlier issues (Beaton-Tukey and Tukey Optimal Influence Function weighting schemes) and multicollinearity issues. We refer to regression using Beaton-Tukey outlier-adjusted weighting and principal components (PCA) analysis as WLRR analysis. Regression modeling using US stocks is referred to as US Expected Returns (USER) and using global stocks as Global Expected Returns (GLER). Regression modeling will use various (M, S, and MM) robust procedures and various (Huber, Hampel, Andrews, Tukey, and Yohai) weighting schemes. The MM methods, using the Tukey and Yohai Optimal Influence Functions, enhance stock selection modeling.

The authors develop variations on Markowitz and Sharpe portfolio optimization techniques, which will illustrate the relative efficiency of individual variables (sales, earnings, book value, dividends, cash flow, forecasted earnings, EP, BP, DP, SP, CP, and FEP) and robust regression-weighted stock selection models.

The authors develop and test the Markowitz-Xu Data Mining Corrections (DMC) procedure and compare it with more recently developed tests. We report statistically significant DMC results. We have significant experience as teachers and practitioners in financial theory, valuation, and financial modeling. We have intimate knowledge of the data available to bridge the theory and application and show how to enhance portfolio returns and maximize terminal wealth.

What Should You Know about the Examples?

This book includes tutorials for you to follow to gain hands-on experience with SAS.

Software Used to Develop the Book's Content

We use SAS 9.3 TS1M0 for Windows 7 Professional 32-bit system to run all the SAS code.

Example Code and Data

You can access the example code and data for this book by linking to its author page at support.sas.com/guerard.

SAS University Edition

This book is compatible with SAS University Edition. If you are using SAS University Edition, then begin here: https://support.sas.com/ue-data .

We Want to Hear from You

Do you have questions about a SAS Press book that you are reading? Contact us at saspress@sas.com.

SAS Press books are written *by* SAS Users *for* SAS Users. Please visit sas.com/books to sign up to request information on how to become a SAS Press author.

We welcome your participation in the development of new books and your feedback on SAS Press books that you are using. Please visit sas.com/books to sign up to review a book

Learn about new books and exclusive discounts. Sign up for our new books mailing list today at https://support.sas.com/en/books/subscribe-books.html.

Learn more about these authors by visiting their author pages, where you can download free book excerpts, access example code and data, read the latest reviews, get updates, and more:
support.sas.com/guerard
support.sas.com/xu
support.sas.com/zwang

Chapter 1: Why Do We Invest?

1.1 Introduction

Consumers balance their current needs and wants of consumption – spending on food, housing, and other expenses – with their desires for future consumption, such as educational expenses, vacations, or buying a long-desired sports car at age 65. People save from their current income to invest in assets that will grow over time so that they can consume more in the future. The purpose of this book is to show readers how use SAS to enhance their wealth.

The total US household net wealth has grown from net wealth of $87 trillion at the end of 2016 to $100 trillion, as reported by the US government (Board of Governors of the Federal Reserve System (2018)), see Torry (2018). The stock market's 20-plus percent return in 2017 is a significant contribution to this jump of wealth. Out of the $100 trillion, $28 trillion are earmarked as retirement assets, which have various tax-favored treatments. The Census report shows that at least 75 percent of households had at least $50,000 in net assets at the end of 2011. Those assets in a household's portfolio might consist of cash, bond, stocks, real estate and so on. These different assets provide different return and risk characteristics. The Federal Reserve reported that the median financial asset value is $200,000 for people 40 years older.

Everybody is a financial asset investor, either passively or actively. If you are the owner of a house, you are an investor in real-estate. It is often stated that your real estate investment is the largest investment a family will make. A recent examination of the Case-Shiller Housing Index reveals that the US housing market has reached an all-time high, as measured from 1970. If you have 401k accounts, you are likely a passive investor. If you have a brokerage account and trade a lot, you are active investor. Even the pension amount that you receive at retirement depends on the market performance. Figure 1.1 provides the cumulative wealth of an investor who invested $100 at beginning of 1928 in either cash, bond, or US stock market.

Figure 1.1: Cumulative Wealth of $100 Starting in 1928

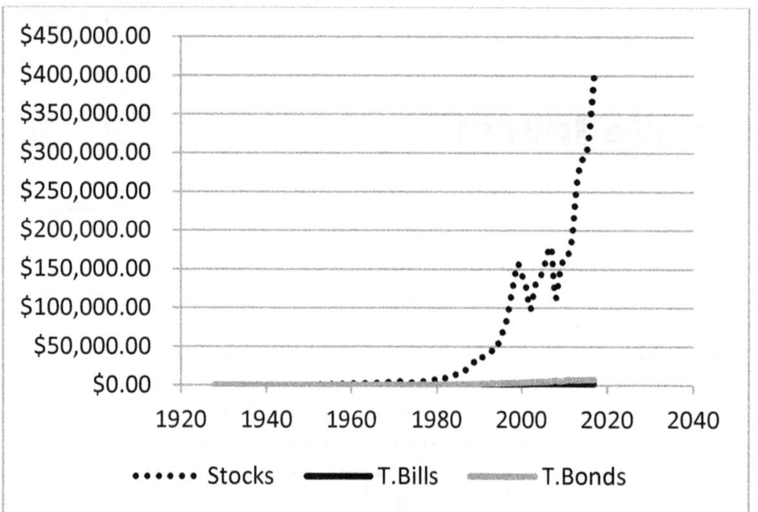

As Figure 1.1 shows, stocks have outperformed bonds, and bonds have outperformed cash. One hundred US dollars at the beginning of 1928 has turned into almost $400,000 at the end of 2017, a return of 400,000 percent for a holding period of 90 years, despite the fact that this period included the Great Depression of 1930 and the 2008 recession. Wealth accumulation does not grow monotonously. The stock market lost about 40 percent of its value recently in year 2001–2002, and year 2008. We use these three types asset as example because they have the longest return history. The S&P/Case-Shiller US National Home Price Index published by the Federal Reserve Bank of St. Louis, a popular index for the real-estate asset class, was started in 1987.

The American people have recognized the earnings power and the risk of stock. The Organization for Economic Co-operation and Development (OCED) reports that only 13.5 percent of US household's financial asset is cash. Xu (2015) reported that young Americans invest more than 80 percent in stocks in their 401k accounts.

The wealth creation power of equity is not limited to the US. Figure 1.2 shows wealth accumulation by investing in international stocks. One hundred dollars invested at the beginning of 1970 in the European market turned into $10,000 at the end of 2017, a return of 10,000 percent with a holding period of 47 years.

Figure 1.2: Cumulative Wealth of $100 Starting in 1970

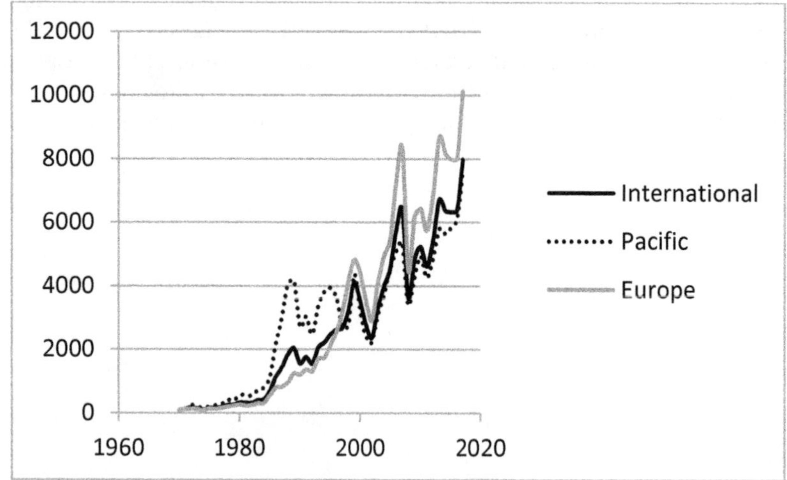

1.2 Assumptions

All writers have beliefs, even prejudiced views. We will disclose our beliefs at the outset. We believe in Active Quantitative Management using the portfolio selection, construction, and management techniques of Harry Markowitz, William (Bill) Sharpe, Jan Mossin, John Blin, Henry Latane, Martin Gruber and Ed Elton, Barr Rosenberg, Haim Levy, and the investment professionals at Factset, FIS, and Axioma. We believe that the empirical evidence of the past 30 years suggests that financial anomalies were identified, have persisted, and most likely will persist into the coming decade.

We believe the benchmarks established by Markowitz, Sharpe and Blin are still relevant, and difficult to beat. New data, better computational power, and enhanced statistical analyses are shown and discussed in this text. We believe that new data, also known as "Big Data," will enhance returns in the future, but the enhancements will be more in the 15–20% range, rather than doubling existing excess returns. We present models, updated analyses, and evidence to show that robust regression, as we estimated in financial models nearly 30 years ago, still works today.

Earnings forecasting can be used to greatly enhance portfolio returns in US, and particularly, non-US markets. Financial models, when properly developed and tested with proper transactions cost, work about 75 percent of the time. The models produce statistically significant excess returns in the years that the models win, but ONLY if the models are used religiously, 100 percent of the time and asset owners fully invest the Mean-Variance weights (or Equal Active Weights, plus or minus the benchmark, at least two percent active weighting, which generally produce lower Sharpe ratios).

Henry Latane and Harry Markowitz taught us in 1959 that to maximize the Geometric Mean maximizes the utility of final wealth; achieving the greatest level of terminal wealth, in the shortest time possible. Henry Latane, in his UNC Portfolio Analysis doctoral seminar, often joked that the Efficient Markets hypothesis only said that the average investor only earned an average return, adjusted for risk. Latane asked, "Who wants to be average?" We believe that smart people, with good databases, can enhance returns about 1–2 percent, annualized, adjusted for risk and risk premiums accepted (knowing or unknowingly incurred). The authors detest closet benchmark-huggers.

1.3 Annualized Return

Cumulative wealth is the product of multi-period returns. If an asset has returns R_t, for t=1,...,T, where t is the period, and T is the total periods, then cumulative wealth is calculated as

$$W_T = W_{T-1} * (1 + R_T) = W_0 \cdot \prod_{t=1}^{T} (1 + R_t) = W_0 \cdot (1 + R_{HP}) \tag{1}$$

where R_{HP} is the holding period return. Even though the holding period returns are correct numbers to capture the wealth change, it is difficult to compare the merit of asset returns with different holding periods. In our case, we have US return data going back to 1926 and MSCI international data going back to 1970. We need a number for merit to compare the returns of US stock and international stock, one with a 90-year holding period, and one with a 47-year holding period. To facilitate comparison, financial reports often use a one-period return instead of holding period return. This one-period return is called the *annualized return*, which is calculated from the holding period return. If this calculated return is realized for every period, then the cumulative wealth will be equal with the wealth generated by the actual holding period return.

$$W_T = W_0 * \prod_{t=1}^{T} (1 + R_t) = W_0 * (1 + R_g)^T$$

From this equality, we derive the formula

$$R_g = \sqrt[\frac{1}{T}]{\prod_{t=1}^{T} (1 + R_t)} - 1 \tag{2}$$

Table 1.1 shows that it is much easier to compare the annualized return than to compare the holding period returns. For the long period of 1928–2017, US stocks outperformed bonds by 4.77 percent a year. For the shorter period of 1970–2017, US stocks outperformed bonds by 3.36 percentage points a year. For the same period, the US stocks outperformed the Europe stocks slightly by only 30 basis points per year. The European stocks outperformed the Pacific stocks by 65 basis points.

Table 1.1: Annualized Return in Percentage

Holding Period	US Stock	US Bond	International	Pacific	Europe
1928-2017	9.65	4.88			
1970-2017	10.45	7.09	9.55	9.44	10.10

1.4 Average Return

Although annualized return is an accurate yearly number to describe the earning power of each asset, from a portfolio's point of view, it lacks the additive property. In other words, the portfolio's annualized return is not the portfolio's weighted annualized return. It is therefore more convenient to work with the simple average return in portfolio management, which is

$$R_a = \frac{\sum_{t=1}^{T} R_t}{T} \quad (3)$$

The average return is often called the arithmetic mean. The annualized return is often called the geometric mean. Table 1.2 shows the arithmetic means of the same five assets shown in Table 1.1.

Table 1.2: Average Return in Percentage

Holding Period	US Stock	US Bond	International	Pacific	Europe
1928-2017	11.53	5.15			
1970-2017	11.83	7.50	11.74	12.92	12.20

Notice that the arithmetic means reported in Table 1.2 are always higher than the geometric means reported in Table 1.1. The Pacific stock is the best-performing asset according the arithmetic mean, while US stock is the best-performing asset in reality according to Table 1.1. The arithmetic mean ignores the variability of returns whereas geometric mean couples the variability of return too tightly. The geometric mean is roughly the arithmetic mean subtracted by half the variability

$$R_g \approx R_a - 0.5 * \sigma^2 \quad (4)$$

1.5 Expected Return

We can use SAS to calculate the means returns, return variabilities, and return correlations of the assets from Tables 1.1 and 1.2, and report the results in Output 1.1.

Program 1.1: Correlations of Global Market

```
proc corr data = global_assets_annual_returns;
     var USStocks USBonds Pacific European;
run;
/*Data Source: SBBI(2017)*/
```

Output 1.1: Results

The SAS System

The CORR Procedure

4 Variables: USStocks USBonds Pacific European

Simple Statistics

Variable	N	Mean	Std Dev	Sum	Minimum	Maximum
USStocks	48	11.83	16.79	567.72	-36.55	37.20

Simple Statistics

Variable	N	Mean	Std Dev	Sum	Minimum	Maximum
USBonds	48	7.50	9.55	360.09	-11.12	32.81
Pacific	48	12.92	29.19	620.35	-36.17	107.55
European	48	12.20	21.45	585.72	-46.08	79.79

Pearson Correlation Coefficients, N = 48
Prob > |r| under H0: Rho=0

	USStocks	USBonds	Pacific	European
USStocks	1.00000	-0.02015	0.44319	0.75717
		0.8919	0.0016	<.0001
USBonds	-0.02015	1.00000	-0.18788	-0.05029
	0.8919		0.2010	0.7343
Pacific	0.44319	-0.18788	1.00000	0.58032
	0.0016	0.2010		<.0001
European	0.75717	-0.05029	0.58032	1.00000
	<.0001	0.7343	<.0001	

Asset return is modeled as random variable X in the modern portfolio theory. The expected value of this random variable $E(X)$ is called expected return. Expected return is often notated using Greek symbol μ in financial literature convention. The variance of this random variable $V(X)$, or the standard deviation $\sigma_x = \sqrt{V(X)}$, is a measurement of risk in financial literature.

Assume that there are n investable assets with expected return vector $\mu' = (\mu_1, \mu_2, \ldots, \mu_n)$ and variance covariance matrix $C = \begin{pmatrix} c_{11} & \cdots & c_{1n} \\ \cdot & \cdot & \cdot \\ c_{n1} & \cdots & c_{nn} \end{pmatrix}$, where c_{ij} is the covariance of asset i with asset j,

If portfolio P has weight $w' = (w_1, w_2, \ldots, w_n)$ on assets i=1,2,…,n. Then the portfolio's expected return is

$$\mu_p = \sum_{i=1}^{n} w_i \mu_i \tag{5}$$

And the portfolio's variance is

$$V_p = \sum_{i=1}^{n} \sum_{j=1}^{n} w_i w_j c_{ij} \tag{6}$$

1.6 Efficient Portfolio

The decision variables in portfolio theory are portfolio weights. Markowitz (1952, 1959) created Modern Portfolio Theory, often denoted as MPT, and stipulated that portfolio weights should be chosen such that the portfolio is mean-variance efficient. In other words, no other portfolio has higher expected returns with the same risk and no other portfolio has lower risk with the same expected return. Mean-variance efficient portfolios are also called *efficient frontier*. The efficient frontier can be found by quadratic programming

$$\min_w V_p \tag{7a}$$

Such that

$$\mu_p = E_p \tag{7b}$$

$$Aw = b \tag{7c}$$

$$w \geq 0 \tag{7d}$$

where constraint (7b) is portfolio's expected return and (7d) is no-short selling constraint, i.e. every weight w_i must be nonnegative, and constraint (7c) is used to make sure the portfolio has the desired characteristics like industry exposures and factor exposures. A is an $m \times n$ matrix, and b is an m component vector. There is no analytic solution in general. In textbook portfolio theory, the only linear constraint of (7c) is usually budget constraint

$$\sum_{i=1}^{n} w_i = 1 \tag{8}$$

and the short-selling constraint (7d) is often ignored. In this case,

the efficient portfolio can be found by the following unconstrained optimization problem

$$\min_w V_p - \lambda_\mu \left(\sum_{i=1}^{n} w_i \mu_i - E_p\right) - \eta\left(\sum_{i=1}^{n} w_i - 1\right) \tag{9}$$

where Lagrange multiplier λ_μ is called risk return trade-off parameter in financial literature.

By taking the derivative with respect to all weight variable w_i, the first order conditions of (9) in matrix form is

$$2Cw = \lambda_\mu \mu + \eta \ell$$

where ℓ is the vector of ones, i.e. $\ell' = (1,1,...,1)$. This implies the optimal portfolio weight has the general form

$$w = 0.5\lambda_\mu C^{-1}\mu + 0.5\eta C^{-1}\ell \tag{10}$$

Together with expected return constraint (7b) and budget constraint (8), the efficient portfolio is

$$w = \frac{E_p \ell' C^{-1}\ell - \mu' C^{-1}\ell}{\mu' C^{-1}\mu \ell' C^{-1}\ell - (\mu' C^{-1}\ell)^2} C^{-1}\mu + \frac{\mu' C^{-1}\mu - E_p \mu' C^{-1}\ell}{\mu' C^{-1}\mu \ell' C^{-1}\ell - (\mu' C^{-1}\ell)^2} C^{-1}\ell \tag{11}$$

The variance of the corresponding portfolio is;

$$V_p = \left(\frac{E_p \ell' C^{-1}\ell - \mu' C^{-1}\ell}{\mu' C^{-1}\mu \ell' C^{-1}\ell - (\mu' C^{-1}\ell)^2}\right)^2 \mu' C^{-1}\mu + \left(\frac{\mu' C^{-1}\mu - E_p \mu' C^{-1}\ell}{\mu' C^{-1}\mu \ell' C^{-1}\ell - (\mu' C^{-1}\ell)^2}\right)^2 \ell C^{-1}\ell$$

$$+2\frac{E_p \ell' C^{-1}\ell - \mu' C^{-1}\ell}{\mu' C^{-1}\mu \ell' C^{-1}\ell - (\mu' C^{-1}\ell)^2}\frac{\mu' C^{-1}\mu - E_p \mu' C^{-1}\ell}{\mu' C^{-1}\mu \ell' C^{-1}\ell - (\mu' C^{-1}\ell)^2}\mu' C^{-1}\ell \qquad (12)$$

The simple constrained portfolio optimization problem does have an analytical solution.

1.7 Minimum Variance Portfolio

Let us study a special portfolio with the expected return set to

$$E_{\min} = \frac{\mu' C^{-1}\ell}{\ell' C^{-1}\ell} \qquad (13a)$$

The corresponding portfolio weight vector is

$$w_{\min} = \frac{C^{-1}\ell}{\ell' C^{-1}\ell} \qquad (13b)$$

The variance of this portfolio is

$$V_{\min} = \frac{1}{\ell' C^{-1}\ell} \qquad (13c)$$

Portfolio $w_{\min}$ is called the minimum variance portfolio. It achieves the minimum risk among all the portfolio combinations.

If we choose the portfolio's expected return to be

$$E_p = E_m = \frac{\mu' C^{-1}\mu}{\mu' C^{-1}\ell} \qquad (14)$$

Then the corresponding efficient portfolio weight is

$$w_m = \frac{C^{-1}\mu}{\mu' C^{-1}\ell} \qquad (15)$$

And the variance of this portfolio is

$$V_{\mathrm{m}} = \frac{\mu' C^{-1}\mu}{(\mu' C^{-1}\ell)^2} \qquad (16)$$

1.8 Market Portfolio

The portfolio w_m is called the market portfolio. Equation (11) is a special case of the two-fund theorem, which states that all efficient portfolios are a linear combination of two basic efficient portfolios. Here the two basic efficient portfolios are the minimum variance portfolio and market portfolio. If an investor can borrow and lend at a risk-free rate, then the minimum variance portfolio is the portfolio composed of 100 percent risk-free assets. The two-fund theorem becomes the two-fund separation theory of Tobin (1958).

We can use the sample average return as expected returns vector and the sample covariance as variance covariance matrix to generate global efficient frontier as depicted in Figure 1.3.

Figure 1.3: Global Risk-Return Tradeoff (Efficient Frontier): 1970-2017

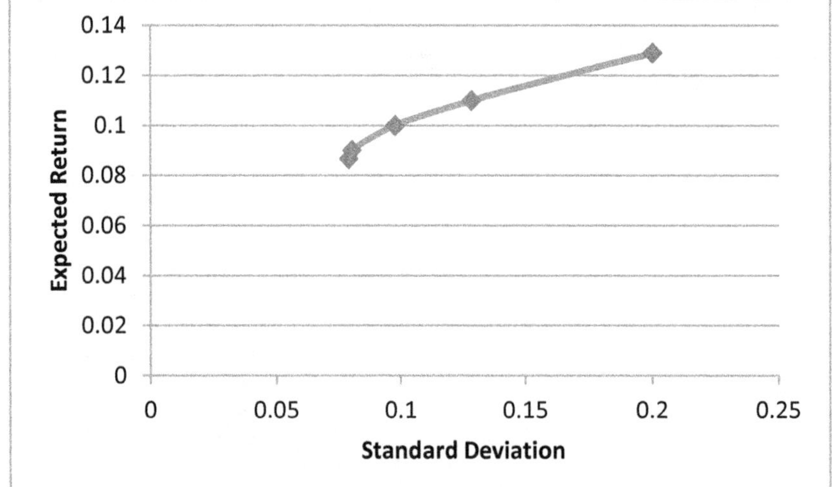

The minimum variance portfolio and market portfolio with four asset classes consisting of US stock, US Bond, Pacific Stock, and Europe Stock is presented in Table 1.3.

Table 1.3: Global Minimum Variance and Market Portfolios

Asset Classes	$w_{\min}$	w_m	w_{gm}
US Stock	0.26	0.20	0.79
US Bond	0.66	0.75	0.00
Pacific	0.09	0.08	0.19
Europe	-0.02	-0.03	0.02
Expected Return	0.09	0.09	0.12
Standard Deviation	0.08	0.08	0.16

In real-life portfolio management, some form of nonnegative constraint (7d) and affine constraint (7c) is always present.

Markowitz (1959) invented a fast and efficient way – the critical line algorithm – to find all of the efficient portfolios satisfied by the general form of constraint (7c) and (7d) with expected return E_p as a parameter. Block et al (1993) used the critical line algorithm to run hundreds of simulations in the 1990s. The general form of constraint (7c) can handle inequality constraints too by adding auxiliary slack variables.

As expected, the unconstrained efficient frontier dominates the constrained efficient. The difference is small in this four-asset class case. This is foreseen by Table 1.3 in which both the minimum variance portfolio and market portfolio are short selling only two and three percent of Europe stock. Two or three percentage differences in weights does not change the portfolio's mean and risk significantly.

We have used sample means as expected returns and the sample variance-covariance matrix as input to generate efficient portfolios arising from the optimization problem of general form (7). In Table 1.3 we report a geometric mean maximizing portfolio w_{gm}, which is 79 percent US stock, 19 percent Pacific stock, and two percent European stock. This is the second portfolio from the right reported on the constrained Efficient Frontier in Figure 1.4. Portfolio w_{gm} achieves the maximum geometric mean.

Figure 1.4: Comparison of Constrained vs Unconstrained Risk-tradeoff curves.

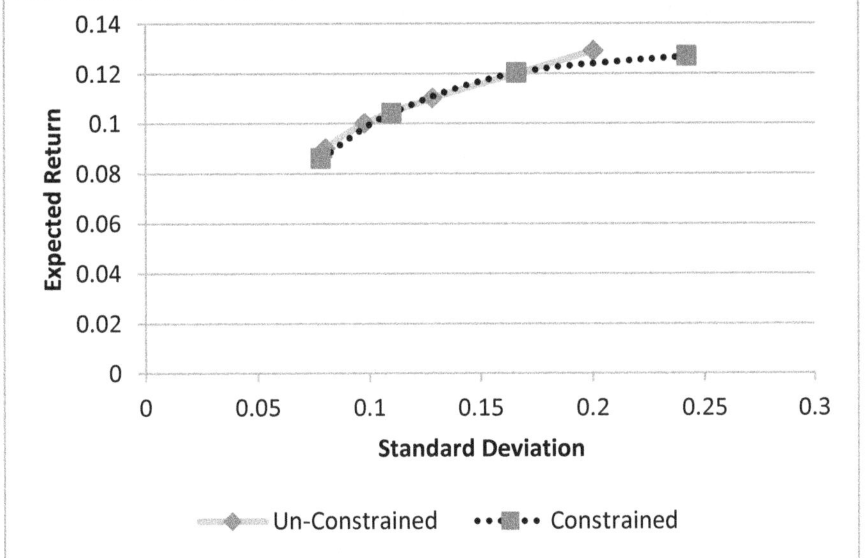

1.9 Portfolio Optimization

The mean-variance portfolio analysis enables investors to choose the best portfolio to suit their risk tolerance. Retirees do not have to invest all money in bond. The minimum variance portfolio $w_{\min}$, with two-thirds of the fund in bonds and one-third of the fund in stocks, is as low risk as buying 100 percent US bonds while making one hundred basis points more return. Young people who are ages 35 and younger and who would like to grow their funds should invest in geometric mean maximizing portfolio w_{gm}. Figure 1.5 shows the cumulative effect of optimal portfolio combinations.

Figure 1.5: Maximizing the Geometric Mean and Terminal Wealth

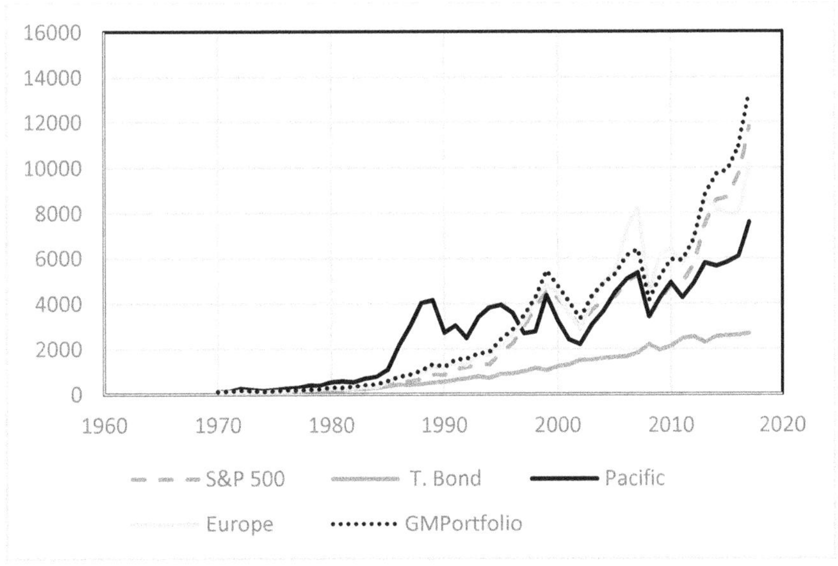

As shown in Figure 1.5, $100 starting in 1970 turns into $13,350.68 for investing in the gm-maximizing portfolio while US stock turns into $11,799.48. This is a great enhancement in terminal wealth!

This book explains how to build expected returns of a thousand securities as portfolio optimization input. The past average return contains only part of the information related to future expected return. There is also future-related information in the fundamental variables, see Block et al (1993) and later chapters of this book. We will introduce financial statements and financial ratios in Chapter 2, and then teach how to build expected return models in Chapter 4. Modern Markowitz-based portfolio construction models are explored in Chapter 5.

The goal of active management is to beat market portfolio by a couple of hundred basis points on an annual basis. Figure 1.6 shows the astounding results of beating the market by one hundred basis points. This book shows that it is possible by deploying advanced statistical tools and disciplined scientific portfolio optimization.

Figure 1.6: The Cumulative Effect of Beating the Market

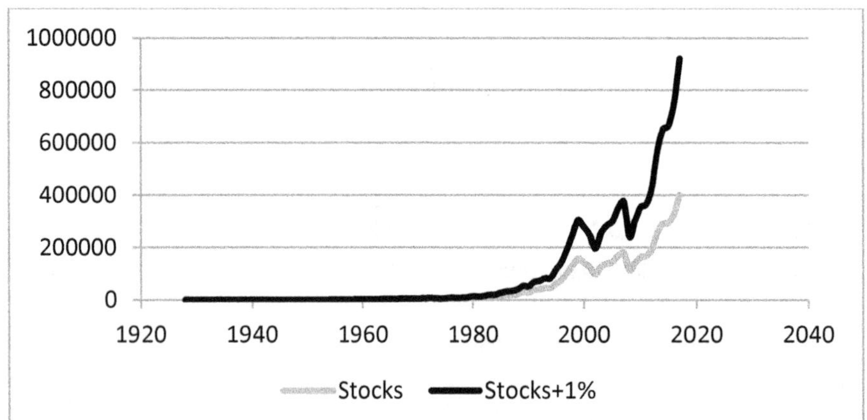

1.10 Summary and Conclusions

In this chapter we introduced the reader to the concept that current savings can be invested into stocks and bonds that can enhance terminal wealth. This is a risk-return concept that is extremely important for investors. Stocks have produced more returns than bonds over the past time periods from 1928 – 2017 and 1970-2017 because their risk, as measured by the standard deviation is greater. The risk-return trade-off does change over time, but if an investor has a 30-year investment horizon, then stocks will be preferred to bonds to maximize terminal wealth. If an investor can earn more than one percent above the market return, then terminal wealth is greatly increased. The purpose of this book is to show readers how use SAS to enhance their wealth.

Chapter 2: An Introduction to Financial Statement Analysis

2.1 Introduction

This chapter on financial statement analysis introduces the reader to the income statement, the balance sheet, the cash flow statement, and ratio analysis. We seek to acquaint the reader with accounting and financial terminology and to understand the importance of corporate earnings, earnings per share, and earnings per share forecasting in finance and investments. We assume that the reader has never seen a set of corporate financial statements and has never heard of a financial ratio.

The corporation is the major institution for private capital formation in our economy. The corporate firm acquires funds from many different sources to purchase or hire economic resources, which are then used to produce marketable goods and services. Investors in the corporation expect to be rewarded for the use of their funds; they also take losses if the investment fails. The goal of corporate financial management is to maximize stockholder wealth.

Simply put, corporate finance is concerned with how the firm raises funds for its operations, produces goods and services, generates cash flow, and generates returns for its investors. The study of corporation finance deals with the legal arrangement of the corporation, the instruments and institutions through which capital can be raised, the management of the flow of funds through the individual firm, and the methods of dividing the risks and returns among the various contributors of funds.

The purpose of this chapter is to present the reader with tools to analyze the financial state of a corporation, determining from a financial perspective which companies are more likely to fall in value and which companies are more likely to rise in value. Ratios from balance sheets and income statements can be used to predict the likelihood of bankruptcy. We will use several of the financial variables developed in this chapter in constructing efficient portfolios in Chapters 4 and 5. Furthermore, many of the variables are used in Chapter 6 for constructing the Markowitz-Xu Data Mining Corrections (DMC) test. Financial data represents information available to investors and management. How can and should investors use this information? What is the value of the financial information? We specifically address the value of the financial information in chapter 6. Much of the material about the types of businesses and the basis definitions of income statement, balance sheet, and sources and uses of funds statements are adapted from Guerard and Schwartz, *Quantitative Corporate Finance* (2007). That text, used as a supplemental second-year MBA finance text, built upon Schwartz, *Corporate Finance* (1962), and Guerard and Vaught, *The Handbook of Financial Modeling* (1989). The materials in Chapters 2 and 3 are introductory in nature. Our purpose is to acquaint all readers with a solid base knowledge of financial statements and ratios to access the health of firms.

2.2 Types of Businesses

There are three types of businesses: proprietorships, partnerships, and corporations. Economists have recognized these business types since Arthur Stone Dewing discussed these business types in his seminal book, *The Financial Policy of Corporations* (1953). There are a great number of proprietorships in the US. Many more businesses are proprietorships than partnerships or corporations. However, corporations produce the highest net incomes, float more stocks and bonds, and engage in more financing activities than proprietorships and partnerships, see Guerard and Schwartz (2007).[1] Thus, corporations produce the vast majority of net income in the US economy.

Proprietorships

The organization of the single proprietorship involves little legal formality. The owner and the business firm are legally one. No special legal permission is required by the state to set up a sole proprietorship. Proprietors have legal title to the assets of the business. Proprietors personally assume all debts.

If, in the course of operations, the assets of the business fail to satisfy all of the business liabilities, a proprietor's personal wealth or holdings can be used to help cover the claims of business creditors. Moreover, conversely, the net business assets are subject to the unfulfilled claims of personal creditors. This constitutes the basic rule of "unlimited liability for all debts whether personal or business." The unlimited liability rule actually strengthens the relative credit position of the single proprietor because the proprietor's personal wealth acts as a sort of second guarantee for the safety of the business debts. A major drawback of a proprietorship is that a failing venture might cost an individual not only the funds directly risked in the business, but the rest of the moneys, assets, or wealth he might have reserved for personal use.

The single proprietorship has the advantages of simplicity and direct responsibility in management. It is limited, however, in its sources of ownership capital and in its ability to attract specialized managerial talent.

Partnerships

A partnership is an agreement by two or more individuals to own and run a business jointly. Traditionally, many professional firms such as accounting firms and law firms were organized as partnerships. The agreement can be oral, but in most cases it is in writing to prevent possible subsequent disputes. The usual clauses in a written agreement are fairly well standardized. The partnership has often been defined as a contract of mutual agency. The legal profession repeatedly advises that one should have confidence in the reliability and judgment of the other party before entering a partnership.

The ordinary partner (strictly defined as a general partner unless the general partner is an LLC, a limited liability corporation) has unlimited liability for the partnership's business debts.[2] Thus, if the partnership fails and the firm's assets fail to cover its liabilities, the creditors can seek to recoup their losses from the partners' private assets. Moreover, it is not the duty of the creditors to apportion losses among the partners. They might seek compensation for their claims where they can find it, regardless of any loss-sharing agreement among the partners. If one partner's personal assets are greater, or simply more available, he might well suffer disproportionate losses. A partner who loses more than his agreed share might have a counterclaim against the other partners,[3] but, of course, collection under these circumstances might be delayed considerably.

Forming a partnership does not require any specific permission from the state. Each partner can bind the others to contracts incurred in the normal operations of the business. No matter what formula has been set up for sharing returns out of the company, each partner will be taxed at the appropriate individual income tax rate for all the income realized or imputed to him out of the operations of the firm. Thus, if a partner receives a salary for his services to the partnership, interest on loans to the partnership, and a share of the remaining pro forma profits,[4] he will pay tax on these items. He must pay tax on his total share of partnership profits, whether he draws them out for personal use or reinvests them in the business. The partnership as a firm files a Form 1065 Information return and distributes K-1s to the partners. Its report is made purely for information purposes so that 100 percent of the partnership income can be imputed to the various partners.

Because the partnership rests on a foundation of mutual trust, its dissolution is made very easy. If no specific provision is written into the agreement, any partner can call for dissolution, usually with a required notice of 30 days. Then a procedure called an "accounting" takes place; the assets are liquidated, and each partner is paid his ownership share (equity) according to the agreement. Since the assets upon liquidation are unlikely to bring in their "going concern" value, dissolution usually entails an economic loss. Often the partnership agreement sets a definite liquidation amount to be paid to a partner who requests dissolution, so that the whole going-concern does not need to be liquidated.

Since the partnership is a contract of close personal relationship, a partner's interest in the firm cannot be easily transferred or sold to a third party although the partner's interest can be assigned on death. A new party cannot enter the partnership without the consent of all the partners. If with the consent of the other partners, one partner sells out his interest in favor of someone else, the old partnership is actually legally dissolved, and a new firm is formed. Again, since the partnership is constructed on the basis of personal contracts, the death of any partner dissolves the relationship.[5]

The main advantages of partnerships are flexibility in operation and ease of organization. More managerial talent can be assembled than is possible under the single proprietorship. The partnership can bring together more capital than is likely in the single proprietorship, and it might have a better credit standing relative to that of a small corporation because of the unlimited liability of the partners.

On the other hand, the partnership might engender tensions between its principals and the difficulties of selling a partnership interest and the possible losses from a forced dissolution caused by either the death or withdrawal of one of the partners.[6]

Corporations

The corporation is defined as a fictitious person created by the state. It can engage in certain defined activities, and in pursuing its purposes, it can obtain title to or dispose of property, enter into contracts, and engage agents to work and act for it. Of course, the corporation is not a human being and can act only through humans hired to work for it, yet the legal fiction of the "corporate person" is nevertheless a highly ingenious device. It enables the corporation to take responsibility and liability for legitimate activities which otherwise would be the final liability of the individuals in the organization. Unless something illegal has occurred, people dealing with the corporation must satisfy their claims against the corporation and might not look to those behind the company for ultimate settlement.

The limited liability of the owners is the single most important characteristic of the corporation. Limited liability simply means that in case of failure, the owner or stockholder might lose what she has ventured in the firm, but even if the company cannot pay its creditors in full, the shareholder's personal assets cannot be endangered.[7] Because it possesses limited liability, the modern business corporation provides a method for many people to risk their funds in a distant organization, perhaps with the chance of gain if the business goes well, but with a definite limit of loss if the worst should happen. It is difficult to imagine how large corporations with funds amassed from many individuals could have developed without this provision.

Another important aspect of the corporate form is that its ownership shares are transferable. Unlike the partner, the owner of a share in a corporation has the inherent right to sell or give away his holdings without the consent of the other shareholders.[8] The new holder acquires the same rights and privileges as the original owners. The characteristic of transferability is especially important to the development of modern free enterprise economies since it provides a high degree of capital mobility. Indeed, the ready transferability of corporate shares makes possible the whole intricate structure of capital markets.

The length of life of the corporate firm is independent of its stockholders. If a shareholder dies, her heirs take title to her shares, and the other surviving stockholders have to accept the new owners; there is no legal problem of "succession." The corporation virtually has "perpetual life." Although most corporation charters granted by the states run from twenty to forty years and only a few are perpetual, most charters might be renewed with ease.

The corporation is a stable form of business organization because it can be dissolved only by a majority vote of the stockholders. In contrast, a partnership is dissolved on the death of a partner or on the request of any partner if he does not approve of the way the partnership is run or if he is displeased with his present or potential partners. In exchange for stability, the shareholder sacrifices the greater flexibility and the choice of associates inherent in the partnership. If the shareholder does not care for the other stockholders or if he thinks he can use his capital more productively elsewhere, his only remedy is to sell his shares for what he can get. A dissident corporation shareholder cannot obtain dissolution of the firm.

The single proprietor or the partners take direct responsibility for managing their firm. The owners of the corporation (i.e., the stockholders) have only an indirect role in the actual management of the company. The stockholders have the right to vote to elect a board of directors, who then appoint corporation officers, a President, Secretary, Treasurer, and various vice presidents, to run the business and make the day-to-day decisions for the firm. In a closely held company, where one person or a small, closely connected group holds a majority of the shares, the major stockholders, the Board of Directors, and the management (i.e., the major officers) might all be the same people. In general, a widely held corporation is one where no single stockholder or close group of stockholders holds anywhere near a majority of the shares. Nevertheless, the management (though it usually holds only a small minority of the total shares) is the dominant voice in the corporation's affairs. In theory, the Board of Directors has strong powers. In actuality, the directors are nominated by the management and upon election reappoint the corporate officers. Guerard and Schwartz (2007) examined the Board of Directors of the IBM Corporation, which included university presidents and well-known persons.[9] The stockholders almost invariably vote for the management slate; usually it is the only group running. The voting rights of the stockholders have become in practice a remote, ultimate power that might be used if the firm is badly mismanaged.

Although the lack of influence the average stockholder has on the actual direction of the firm is often considered a drawback of the corporate form, it is not entirely a disadvantage. Most investors are not interested in the ordinary business decisions of the firm. They usually have other concerns, and they would consider it burdensome to devote any considerable time to the operation of the company. The large corporation develops its own management, which after time achieves certain autonomy, might develop a professional outlook, policies, traditions, and very likely a considerable devotion to the affairs and success of the company. Although the remuneration of management might be considerable, the economic interests of the shareholders are well-served by the corporate structure.

In regard to federal income taxation, the status of the corporation is quite unique. Since the corporation is a legal personality in its own right, it is taxed independently on its profits. The stockholders are further taxed at the individual tax rates for the profits that they receive in the form of dividends or at the capital gains rate when reinvested earnings raise the value of shares and these shares are sold at a profit. Because profits have already been taxed once, corporations are described as being subject to "double taxation." In certain circumstances, however, the corporation form might actually offer some tax advantages. For example, if a firm whose owners are in a high-income tax bracket are in a profitable growth stage, it can retain a considerable proportion of its profits for reinvestment in the business. These profits might later be taxed at the lower capital gains rate. In any case, the other favorable aspects of the corporate form decisively outweigh any extra tax burden.[10]

In spite of its weaknesses, the corporate form gives many advantages to its stockholders (owners). If this were not so, it would be hard to account for the growth and economic dominance of the corporation over other forms of private business organization. Essentially, the corporate form permits the pooling of large amounts of capital from many sources (institutional and private) because it does not obligate the individual investor to be active in the affairs of the business nor subject her to worry over personal liability for business debts.

At most periods in our history the publicly held corporation has been able to raise capital with relative ease because over time, investment in a diversified portfolio of corporation shares has shown a comparatively good rate of return. As a corollary, in our economy characterized by the growth of industries requiring heavy capital investment, large-scale operations, and specialized professional managements, the corporation's ability to amass funds and to develop responsible administrative staffs has made it the dominant business form.

The greatest numbers of firms are single proprietorships.[11] In the United States in 2008, there were 30 million business entities that filed tax returns. Most enterprises, over 22.6 million were nonfarm proprietorships, or one-person firms, that produced business receipts (sales) of $1317 billion and net income of $265 billion. Partnerships, some 3.1 million, in 2008, produced business receipts (sales) of $4963 billion and net income of $458 billion. Corporations, totaling 5.85 million, produced business receipts (sales) of $27,266 billion and net income of $984 billion.[12] We stress corporations in this text because corporations produced the most sales and net income, and raise the most capital of the three forms of business entities.

2.3 The Income Statement

Background

The goal of corporate finance is to maximize the stock price of the firm and the market value of the firm. Investors seek to maximize total returns from investing in stocks that include dividends as well as capital gains (which are generated by rising stock prices). The stock price appreciation is the major part of total return for young companies. The stock price of a company should be dependent upon its (recent historical) earnings and its expected earnings. An intelligent investor should be prepared to read, understand, and implement an investment strategy that incorporates financial data of firms.

An investor buys stock, representing (often fractional) ownership of the firm. An investor may receive dividends paid from the earnings of the firm. An investor buys stock in anticipation that the price of the stock will rise in the future, and the stock price appreciation is greater when the company earnings increase by a greater amount than was anticipated by the market. Supply and demand determine the price of common stock. If an investor is buying stock in anticipation of price appreciation, then the seller of the stock believes that its gain is very limited.

An investor anticipating a stock price decline may short the stock, selling it by borrowing from the investor's brokerage, and replacing the stock by purchasing later at a hopefully lower price. We will see variables of forecasted earnings acceleration illustrated in Chapters 5 and 6.

The income statement of a firm reports its sales revenues, costs of its goods sold, and the administrative expenses associated with the sales. Furthermore, the firm pays taxes to Federal, (perhaps) state, and local governments. In this text, we focus on net income, the "bottom line" of the income statement. Net income might be paid out as dividends to stockholders to compensate them for their investment, reinvested in the firm, or used to repurchase previously issued stocks and bonds.[13] Net income will be substantially higher for firms in the future because the US corporate income tax rate was reduced in H.R. 1, the Tax Cuts and Jobs Act of December 20, 2017, from 35 percent to 21 percent, for incomes exceeding 10 million dollars and because firms can now deduct all costs of new investment in the year of service, rather than depreciating them over multiple time periods.

Financial service firms, such as FactSet and Compustat, gather data on corporations for the investment community and produce a similar format for all firms to make it easy to compare companies. The form used by the services breaks out most of the important variables that are interesting for investment analysis. For example, a company's annual report will often lose the depreciation charges in a lumped account such as "manufacturing costs" or "costs of goods sold," and the actual depreciation charges can be obtained only in a footnote or in an obscure part of the report. The financial services show the depreciation charges as a separate item.

Major Items

The following discussion explains the major items appearing in the suggested statement and gives something of their significance.

Sales or Revenues

The sales account shows the total gross revenue received by the firm during the period. It includes sales for cash and for credit, whether or not they were collected at the end of the period. The sales figure should be the net of allowances made to the buyers for returned shipments.

Cost of Goods Sold including Depreciation and Amortization

The direct operating costs are the amounts spent for material and the labor on the goods sold. We refer to these costs as the cost of goods sold. In this chapter, depreciation and amortization charges are included with the traditional costs of goods sold. We will have a discussion of depreciation charges later in the chapter.

Gross Income

This account reflects revenues less the costs of goods sold (COGS), including depreciation and amortization charges.

Earnings before Interest and Taxes (EBIT)

The EBIT, or operating income, is calculated as sales less the cost of goods sold and less selling, general, and administrative (SG&A) expenses. EBIT represents the income of the firm after the book charge for depreciation has been made. After this figure is determined, the effect of many past financial decisions comes into play. The amount of interest that will be paid is based on the amount of interest-bearing debt the firm has incurred, and this influences the profits tax. The amount available for the common stockholders is obtained after dividends on the preferred stock is subtracted. These figures are influenced by decisions on alternate methods of financing the firm. The pros and cons of these decisions make up a large part of the subject of corporation finance.

Depreciation (and Other Noncash Charges)

Noncash charges are items such as depletion, depreciation, or amortization of franchises, patents, etc. Depreciation is usually by far the most important of these items. The depreciation account is the estimated capital (i.e., fixed assets) used up during the year. The depreciation charge is based on the cost – not the present value – of the assets and the schedule of depreciation charges on an asset once set initially cannot be varied except under special circumstances. The level of depreciation charges is important in setting the amount of corporation profit tax due. It is important in reminding the management that not all the returns coming in are income, some must be considered a return of capital, and dividends policies should be set accordingly. H.R. 1, the Tax Cuts and Jobs Act of December 20, 2017, changed depreciation expenses such that businesses can deduct 100 percent of the cost of the eligible property in the year it is placed for service.[14]

However, the annual depreciation charges do not set the amount of fixed assets that will be replaced or new fixed assets purchased. This decision is based on the forecasted future profitability of the replacement or of the new assets. If investment in new fixed assets appears to generate an acceptable rate of return, then available internal funds or other sources of funds will be found. If capital expenditures exceed internally generated cash flows, then new debt or equity must be raised to finance the investment. If the new investment in fixed assets does not exceed the amount of depreciation, then the extra funds can be used for something else,[15] such as retiring long-term debt, repurchasing equity, or paying larger dividends.

Interest

This item represents the interest paid by the company on debt. It is reduced by the amortization of any premium on bonds payable and increased by the amortization of bond discounts. Interest expense is deductible before calculating the corporation income tax.

Corporation Profits Tax

The historic corporate income tax schedule for tax years beginning after December 31, 1992, is shown in Table 2.1.

Table 2.1: Income Brackets and Tax Rates

Taxable Income	Tax Rate
$0 - $50,000	15%
$50,001 - $75,000	25%
$75,001-$100,000	34%
$100,001-$335,300	38%
$335,301-$10,000,000	34%
$10,000,001-$15,000,000	35%
$15,000,001-$18,333,333	38%
> $18,333,334	35%

Source: Small Business, Quickfinder Handbook. www.quickfinder.com

On December 20, 2017, Congress passed a tax reform bill that lowers the US corporate tax rate to 21 percent for income exceeding $10 million.[16] What are the implications for ongoing business concerns? A business owner's market value rises over 20 percent ((1-35%) / (1-21%)), because after-tax income rises by 21.5 percent! Hence, for many business owners, the holidays came early in December 2017.

Earnings After Taxes

The earnings after taxes, or net income, is the earnings of the firm after taxes are paid. Net income is often referred to as the "bottom line" of companies. Net income can be paid to stockholders as dividends or reinvested in capital (assets) investments. Comparisons can be made regarding net income relative to a firm's assets, defined as the return on assets, ROA, sales, or return on sales, ROS, and equity, referred to as the return on equity, ROE. One prefers to see higher values of ROA, ROS, and ROE. The ROA, ROS, and ROE ratios are profitability ratios discussed later in this chapter.

Nonrecurring Losses or Gains (After Taxes)

Nonrecurring losses or gains arise out of transactions such as the sale of fixed assets (buildings, land, or equipment), often associated with discontinued operations or the sale of subsidiaries or investments in securities. Losses can also occur because of natural disasters (floods or fires) or because of liabilities on lawsuits. In any case, these gains or losses do not arise out of the normal operations of the business. These items are given separately because they are special or nonrecurring. If a firm sells a plant or subsidiary at a profit, the earnings for a given year are raised, but the earnings generated by the subsidiary will no longer be available in the future. The tax on nonrecurring gains will probably not be at the regular 35 percent rate, for if the transaction is classified as a capital gain the maximum rate is 28 percent. Whether or not nonrecurring losses are fully deductible for tax purposes depends on the circumstances. It is suggested that when nonrecurring gains or losses occur, they are entered net of taxes (a loss would be reduced if there were regular income tax that could be used as an offset) after the regular part of the income report. Details should be provided in footnotes. In our illustrative statement, there are no nonrecurring items.

Dividends on Preferred Shares

The dividends declared on the preferred shares are subtracted from net profits to obtain the earnings available to the common. Although the preferred dividends are not a legally fixed obligation of the firm, they represent a claim senior (paid first) to any return on common shares, and there can be no calculation of earnings on the common shareholders' investment until they are accounted for. Senior debt claims are paid before junior debt claims are paid and before dividends are paid to stockholders. The preferred dividends are a prior charge from the view of the true residual owners of the firm – the common shareholders.

Earnings Available to Common Stock

This amount represents the accounting profit or earnings accruing to the shareholders of the business after all prior deductions have been made. The term "accounting profits or earnings" is used deliberately. The accounting profits and the "true" or "economic" profits of the firm can differ considerably. (Economic profits is the amount that remains after all imputed costs – such as interest on all the funds used no matter what the source – are subtracted.) However, economists differ on the definition of economic profits, and none of these definitions are easily implemented in practice. Thus, the reported accounting profits serve as a useful available measure of the firm's success. Most stock valuation models rely upon reported earnings. Moreover, under the discipline of the accounting formalities, profits are reported on a sufficiently consistent basis to enable them to be used in the determination of important legal obligations and privileges. But even within the accounting rules there exist legitimate alternative methods of reporting certain expenses and charges, which can cause considerable variation in the operating results of any year.

Dividends on Common Stock

Dividends are shown here as a charge on earnings. In most jurisdictions, however, legally they constitute a charge against surplus, not current earnings, and can be declared as long as there is a sufficient credit balance in the surplus account, even if there are no current profits. Firms can elect to do this. As a practical matter, however, the dividend policies of most firms are conditioned by their current earnings position and not by their retained surplus account, so from the point of view of functional relationships, the order of accounting presented seems quite correct. Common stock dividends are a voluntary distribution of the profits of the firm and not a legal obligation. Their declaration does not reduce the profits of the firm. Thus, dividends are deducted after the earnings on the common are calculated. The profits tax liability of the firm is not affected by either the payment of preferred or common dividends.

Addition to Earned Surplus for the Year

What is left after all dividends are subtracted from the reported profits are the retained earnings, reinvested earnings, or net addition to surplus, for the year. Of course, if expenses exceed revenues, there would be instead an operating loss or deficit for the year. The retained earnings for the period depend on the level of profit and the dividend policy of the company. These in turn are influenced by factors such as the stability and amount of the company's cash flow, the firm's growth prospects, and its need for equity funds either to acquire additional assets or to repay debt.

The retained earnings accumulated over time become the earned surplus of the firm. But, as mentioned earlier, the surplus account indicates only a historical source for the financing of the firm and does not represent an existing fund of cash.

Earnings Per Share and Cash Flow Per Share

The amount earned and paid in dividends on the individual stockholder's share is of more direct importance to him than the total amount earned by the firm. The earnings and trend of dividends on the individual shares in the long run establish their value in the market.

The earnings per share of a company is obtained by dividing the total earnings available to the common stock by the number of shares of stock outstanding. For example, given $100 million in earnings for this year, and 10 million shares of stock outstanding, the earnings per shares are $10 ($100 million/10 million). Clearly, earnings per shares increases as earnings rise, but it can also rise as the number of outstanding shares fall. Shares outstanding fall when a firm buys back its own stock. IBM has been very active in this strategy as we will see in this chapter. Often the earnings per share are shown once, reflecting the regular, recurring income, and again, including extraordinary income items. An additional figure, not always available but often useful, is the cash flow per share. It includes earnings available to the common shareholders plus noncash charges divided by the number of shares. This figure shows the gross funds available per share of stock which can be used to repay debt, acquire assets, and pay dividends. An interesting possibility is to subtract required amortization of debt from the cash flow per share and arrive at the figure of "free" or "disposable" cash flow per share. This figure might prove useful in comparing two firms where earnings are similar but one firm is required to make payments on the principal of its debt.

Example Income Statement

Income statements are easily found for investors and researchers. We can use IBM as an example. The first source an investor or student should examine is the company annual report. The 2016 IBM Annual Report contains 168 pages and reports three years of Consolidated Statement of Earnings (for IBM and its subsidiary companies, on page 84). In the three-year period from 2014–2016, IBM sales revenues fell from $92,793 million to $79,919 million, a 14.1 percent drop. Net income correspondingly fell from $12,022 to $11,872, a 1.3 percent drop. During the same three years, Earnings Per Share, EPS, rose from $11.90 in 2014 to $12.39 in 2016. The reader, at first glance, must be troubled to see Net Income fall and EPS rise. How in the world is this outcome possible? The reader must remember that EPS is net income divided by common stock share outstanding. Common shares outstanding fell from 1010 million shares in 2014 to only 958.714 million shares in 2016, a decrease of 5.4 percent. If IBM repurchases common stock faster than its net income falls (or rises) as we will see throughout this chapter, then its EPS will rise.

In the IBM Annual report of 2016, Chairperson Virginia Rometty, highlighted the growth of the Cognitive Solutions external revenue growth, led by the Solutions Software revenue, including analytics and the security offerings (see pages 35-36 of the Annual Report). IBM is staking a large portion on the Cognitive Solutions Capabilities of artificial intelligence (AI) resource Watson, its use in applications customized by industry, such as the Watson Platform, Watson Health, and Watson Internet of Things, housed with Solutions Software. IBM is betting that the marketplace will respond favorably to its customs solutions growth in revenues and net income.

Google Finance lists the most recent three years of IBM income statement data; four years on AOL Personal Finance, Stock Research; twenty years on the FactSet database; and fifty-six years on the Compustat database. Asset managers and financial researchers prefer to use databases with at least 15–20 years of data. Statistical significant is easier to establish with larger databases. We will focus the readers' attention to the twenty years of FactSet data and the 56 years of the Compustat database.

Table 2.2 is a modified income statement for IBM, for the 1997–2016 period, which is highly useful for various financial analyses. Sales rose from $78,508 million in 1997 to $79,919 million in 2016, an annualized increase of 0.0009, or .09 percent; commonly stated as zero-growth. EBIT, sales less cost of goods sold and selling, general, and administrative expenses rose from $9098 million in 1997 to $12,323 million in 2016, an average annualized increase of 1.53 percent. EPS rose from $3.00 in 1997 to $13.17 in 2016, an average annualized increase of 7.68 percent. IBM repurchased $154.7 million of shares outstanding and the fewer common shares outstanding produced a higher EPS. IBM is not the only firm effectively reducing its shares outstanding by stock repurchases. Sales of IBM have become more "flat-lined," during the 1997–2016 time period, and even more depressed for the 2012–2016 time period. Gross Income and Operating Income rose faster than sales because the gross margin and returns on assets rose. Debt, with a cheaper cost of financing, was increased by IBM and its stockholders enjoyed much higher returns on assets, sales, and equity.

Table 2.2: IBM Income Statements

	DEC '16	DEC '12	DEC '08	DEC '04	DEC '00	DEC '97
All figures in millions of US Dollars except per share items						
Income Statement						
Sales	79,919	104,507	103,630	96,293	88,396	78,508
Cost of Goods Sold (COGS) incl. D&A	42,334	54,936	58,597	60,767	55,972	47,899
Gross Income	**37,585**	**49,571**	**45,033**	**35,526**	**32,424**	**30,609**
SG&A Expense	25,262	28,582	28,389	23,899	20,790	21,511
EBIT (Operating Income)	**12,323**	**20,989**	**16,644**	**11,627**	**11,634**	**9,098**
Nonoperating Income - Net	1,877	2,254	1,423	1,212	617	657
Interest Expense	777	569	673	139	717	728
Unusual Expense - Net	1,077	761	679	672	0	0
Pretax Income	**12,346**	**21,913**	**16,715**	**12,028**	**11,534**	**9,027**
Income Taxes	449	5,298	4,381	3,580	3,441	2,934
Consolidated Net Income	**11,897**	**16,615**	**12,334**	**8,448**	**8,093**	**6,093**
Minority Interest	16	11	0	0	0	0
Net Income	**11,881**	**16,604**	**12,334**	**8,448**	**8,093**	**6,093**
Discontinued Operations	-9	0	0	-18	0	0
Preferred Dividends	0	0	0	0	20	20
Net Income available to Common	11,872	16,604	12,334	8,430	8,073	6,073
Per Share						
EPS (recurring)	13.17	14.83	9.27	5.21	4.44	3.00
EPS (diluted)	12.38	14.37	8.93	4.93	4.44	3.01
Earnings Persistence	82.68	84.77	78.01	90.25	87.54	85.63
Dividends per Share	5.50	3.30	1.90	0.70	0.51	0.39
EBITDA						
EBITDA	16,704	25,665	22,094	16,542	16,629	14,099

Source: FactSet Fundamentals

2.4 The Balance Sheet

The balance sheet is the financial picture of the firm at a precise point of time. The left side of the balance sheet lists, categorizes, and sums the assets, the items of value the firm owns at a given point of time. The top right side lists the debts and other payment obligations that the firm owes; these are the liabilities. The bottom right side lists the net worth, equity, or ownership capital. The fundamental principle of balance sheet is that the left side must be equal to the right side. In other words, the liabilities plus the net worth of the firm must equal the sum of the firm's assets. The balance sheet then presents this equation:

The sum of the Assets = The sum of Liabilities + Stockholder equity

The difference between the liabilities and the assets is the net worth, equity, or ownership capital.

Although the balance sheet or position statement is a useful quantitative picture of a firm's financial position, it is not an exact reflection of the firm's economic worth. The balance sheet is constructed on the basis of formal rules and does not necessarily represent the market value of the firm as either a growing concern, or liquidated (sold off) entirely. Moreover, the balance sheet represents the financial position at the close on the balance sheet date. The assets and liabilities shown are those that accountants have ascertained to exist at that point of time. It is backward-looking in that regard and might not accurately reflect what the assets or liabilities are worth today or are expected to be worth tomorrow. In plain English, balance sheet data records all assets and liabilities at cost, not at their market value. Imagine the market value of inventories of five-to-seven year old computers (with virtually no market value). Since a business is always changing, the balance sheet never represents more than an approximate financial truth.

Several points are provided by the balance sheet equation, which holds that total assets = total liabilities + stockholder equity (or total debits = total credits). Therefore, if the firm increases its total assets (buys more goods, acquires new equipment, increases the money owed it by extending more sales credit), it follows that either the liability or stockholder equity (ownership accounts) must also increase to balance the rise in assets. The firm can increase the amount that it owes its suppliers, borrow from the banks, float a mortgage bond issue, or it can increase its net worth by floating additional common stock or retaining additional earnings in the business. The issuance of debt increases the firm's interest expense, which it must pay to remain in business, but debt issuance does not possibly dilute the ownership and voting interests of management, inherent in an equity issuance. The problem of whether (or not) to acquire additional assets and the related question of choosing the best source out of which to finance the additional assets are central areas of corporate financial decision making. That is, management should finance the expansion of assets by pursuing the debt or equity strategy that maximizes the stock price of the firm.

Assets

The assets that the nonfinancial firm may acquire or own are usually broken down into three major categories: current assets, fixed assets, and other assets. The current assets and the fixed assets are usually much larger than the other assets. The Assets are usually listed in the order of liquidity. Liquidity refers to the ease with which an asset can be sold and turned into cash. All assets can be sold, but how quickly and at what discount? Cash is the ultimate current asset. Receivables can generally be turned into cash quickly. Inventories, as we have discussed with old computers, might not be worth their costs. The quicker they can be sold, and the smaller the discount required to make the sale, the more liquid the assets are. Hence, the company's current assets are typically more liquid than their plant and equipment, which might be difficult to sell even with deep discounts.

Current Assets

The current assets consist of cash, items that in the normal course of business will be turned into cash within a short time, defined as one year, and prepaid items that will be used up in the operation of the business within the year. The three largest accounts making up the current assets are usually in cash, receivables, and inventories. Current assets make up a great part of the firm's liquidity, its ability to meet its maturing short-term obligations. One must remember that the failure to cover one's maturing short-term obligations can lead to bankruptcy.

Cash

Cash is the sum of the cash on hand held out to make change and provide petty cash funds, and the deposits in the bank. IOUs, advances to employees, etc., are not counted in the cash account.

Receivables

The receivables are amounts due to the firm from customers who have bought on credit. They are often separated into accounts receivable and notes receivable. An account receivable is the usual way credit is given in American business practice. It simply means that the buyer of the goods is charged for his purchases on the books of the seller – no other "formal" legal evidence is prepared. If a note receivable is used, the purchaser of the goods has signed a promissory note in favor of the seller. A note, except in certain lines of business where they are customary, is generally required only of customers with weaker credit ratings or those who are already overdue on their accounts.

An account called reserve for bad debts or allowance for doubtful accounts is generally subtracted from the receivables. This is called a valuation reserve; it is an attempt to estimate the amount of receivables that might turn out to be uncollectible. The receivables minus this reserve, the net receivables, is counted as an asset on the balance sheet. The receivables do not normally contain advances or loans that are not part of the normal business or sales operations of the firm. Loans to officers or employees or advances to subsidiaries are generally included in the other assets.

Inventories

Inventories are items making up the finished stock in trade of the business and the raw materials which in a manufacturing firm will in due course be turned into the finished products. In a mercantile or distributing firm the inventory consists basically of "finished goods," that is, items that the company does not have to process further. In manufacturing companies, the inventory divides into three categories: raw materials, work in process, and finished goods. If we consider the current assets from the "flow of funds" aspect, that is, how close they are to being turned into cash, cash will be listed first, of course, next receivables (representing generally sales made but not yet collected), and then inventory. Obviously, for the going concern, finished goods are more current or liquid than work in process, and work in process more so than raw materials. The relative composition of the inventory can become a matter of importance – sometimes unfortunately overlooked – in analyzing the current credit position of a manufacturing firm.

A problem in presenting inventory values on the balance sheet is to keep separate the amount properly ascribed to supplies. Supplies are not part of the normal stock in trade nor are they processed directly into finished goods. Stationery and stenographic supplies, coal used for heat or power, and wrapping material used in a store, are obviously supplies. It is difficult, however, to classify packaging items that are a distinctive part of the final product in a manufacturing concern, or the coal pile of a steel plant. In general, an item that is an integral part of the final product is part of the raw material inventory, whereas items used in corollary functions are supplies. Supplies are usually placed with the miscellaneous current assets; like the prepaid expenses, they represent expenditures made currently, which save outlays in the future.

Valuation of the inventory is an additional problem with which the accountants must wrestle. The usual rule of valuation is "cost or market, whichever is lower." This rule gives a conservative value to the inventory. The inventory is marked down in value if prices have declined since the items were purchased; but if prices have risen, items are valued at their cost to the firm.[17] Firms can now make a choice between the rule of "first-in, first-out" (FIFO) and "last-in, first-out" (LIFO) as methods of inventory valuation.

Under FIFO (which most firms still use) it is considered (whether physically true or not) that sales have been made of the older items, and that the items most recently manufactured or purchased compose the inventory. Conversely, if LIFO is used to value the inventory, then the new items coming in are considered to enter the cost of goods sold, and the cost of the older stock sets the value of the inventory. Under the first method, FIFO, the value given the inventory on the balance sheet is meaningful, but the cost-of-goods-sold figure used on the income statement might not truly represent current economic costs if price levels have been changing rapidly. Under FIFO, the accounting figure for cost-of-goods-sold tends to lag behind price level changes, so that reported accounting profits are large on an upturn and decrease rapidly (or turn into reported losses) on a downturn in prices.[18]

On the other hand, LIFO reduces the lag in the accounting for cost-of-goods-sold when price level changes, thus modifying the swing of reported accounting profits during the trade cycle. The LIFO method of inventory valuation, however, tends to develop an inventory figure on the balance sheet that might not be at all representative of any current cost or price levels. The asset value of the inventory can become more and more fictitious or meaningless as time passes. In fact, more large companies surveyed annually by the American Institute of CPAs (AICPA) had switched to LIFO by 1981 (Horngren (1984)). In 1994, a corresponding AICPA survey had 351 large firms using LIFO, with 186 firms using LIFO for 50 percent or more of their inventories.[19] The larger the firm's inventory balance, the greater the potential tax savings for LIFO companies. Moreover, in defense of FIFO, any distortion it produces on the profit and loss statement is not very great for firms that turn over their inventory rapidly, i.e., for firms whose stock is replaced rapidly in relation to their sales. In fact, one of the earlier studies of market efficiency by Sunder (1973) found no significant excess returns for firms switching from FIFO to LIFO in inflationary times (from the date of the change), whereas firms switching from LIFO to FIFO experienced a drop in excess returns over 8 percent.

Miscellaneous Current Assets

Other current assets besides cash, receivables, and inventories are accruals, prepaid expenses, and temporary investments. Accrued items are amounts that the firm has earned over the accounting period but which are not yet collectible or legally due. For example, a firm might have earned interest on a note receivable given to it in the past even though the note is not yet due. The proportionate amount of interest earned on the note from the time it was issued to the date of the balance sheet is called accrued interest, and under modern accounting procedures, it is brought on to the books as an asset.

Prepaid expenses are amounts the company has paid in advance for services still to be rendered. The company might have paid part of its rent in advance or paid in on an advertising campaign that is yet to get underway. Until the service is rendered, the prepayment is properly considered an asset (i.e., something of value due the firm). When the service is rendered, the proportionate share of the prepaid item is charged off as an expense. Temporary investments are holdings of highly marketable and liquid securities representing the investment of temporary excess cash balances. If these are to be classified as a current asset, the firm must intend eventually to use these funds in current operations. If, however, the securities are to be sold to finance the purchase of fixed assets or to cover some long-term obligation, such holdings are more correctly grouped with the other assets or miscellaneous assets.

Fixed Assets

The fixed assets and the current assets are the two important asset classes. The fixed assets are items from which the funds invested are recovered over a relatively longer period than those invested in the current assets. Long-term receivables are included in fixed assets. The fixed assets are also called capital assets, capital equipment, or the fixed plant and equipment of the firm.

The annual balance sheet usually lists a Property, Plant & Equipment-Net figure, composed of Property, Plant & Equipment-Gross, less accumulated depreciation. Almost all fixed assets except land are depreciable. Mineral deposits or reserves and standing timber are considered depletable, and are important for mining and petroleum firms. In determining the value of a fixed asset, we must remember that their economic life is not unlimited, that eventually they will wear out or otherwise prove economically useless in their present employment.

The accounting reports allow for the loss of value on fixed assets through the passage of time by setting up a reserve for depreciation, allowance for depreciation, or accumulated depreciation account. Every fiscal period, a previously determined amount is set up as the current charge for depreciation, and is subtracted as an expense on the income statement. The matching credit is placed in the allowance for depreciation account, where it accumulates along with the entries from previous periods until (1) the allowance for depreciation equals the depreciable value (original cost less estimated scrap value) of the asset or (2) the asset is sold, lost, or destroyed. On the balance sheet, the allowance for depreciation constitutes a valuation account or reserve; the historically accumulated depreciation is subtracted from the original acquisition cost of the fixed assets, and the balance, called net fixed assets, is added into the sum of the total assets.

The problem of making adequate allowance for depreciation and determining the periodic depreciation charge has caused considerable difficulty for accountants. A commonly used depreciation method presented to the public on the reported books is the "straight-line" method. This technique is quite simple, which explains, among other reasons, why it is popular. The probable useful life of the asset is estimated, the estimated scrap value of the asset is deducted from its original cost in order to obtain its depreciable value, and the depreciable value divided by the estimated life gives the yearly depreciation charge. The depreciation charge deducted as an expense remains the same year after year though the net book value of the asset is constantly reduced. For example, assume that a warehouse has an estimated life of 20 years, an original cost of $225,000,000 and a $25,000,000 estimated scrap value. The yearly charge will be $10,000,000 per annum or 5 percent of its depreciable value. At the end of ten years, the accumulated allowance for depreciation will be $100,000,000 and the net asset value of the warehouse will be carried at $125,000,000.

Although the straight-line method is popular for public accounting reports, a constant depreciation charge does not reflect the fact that for most fixed assets the loss in economic value is higher in the earlier periods of use. Traditionally, the Internal Revenue Service allowed firms to adopt alternative, accelerated depreciation policies, such as the Accelerated Cost Recovery System (ACRS), instituted in 1981 and modified by the Tax Reform Act of 1986). Most industrial equipment was classified as having a seven-year life in ACRS, which allowed firms to depreciate annually 14.29 percent, 24.49 percent, 17.49 percent, 12.49 percent, 8.93 percent, 8.93 percent, 8.93 percent, and 4.45 percent, respectively, in years one through eight. The great tax advantage of the accelerated depreciation was that it allowed the company to defer some of its income tax liabilities to the future, thus providing a greater current after tax cash flow. Accelerated depreciation was a significant tax saving in present value terms. Very recently, H.R. 1, the Tax Cuts and Jobs Act of December 20, 2017, changed depreciate expenses such that businesses can deduct 100 percent of the cost of the eligible property in the year it is placed for service.

Depreciation allowances are generally based on the original acquisition cost of the fixed asset to the firm. Any subsequent change in the value of the fixed asset – for example, through price level changes – is generally not reflected in the value of the asset on the books nor in the allowable depreciation rate. The depreciation rate is set by the original purchase price and is not changed for the new price level that might exist currently.[20] Thus, even if the funds released to the business by its depreciation allowances were actually segregated (which they are not) for the replacement of fixed assets, they would not prove adequate if the replacement or reproduction cost of these assets had gone up in the meantime.[21] Many authorities have argued for changes in the basic accounting system that would adjust depreciation rates to movements of the general price level. While there is much in favor of such a reform, many practical difficulties stand in the way. Moreover, many accountants and public finance experts argue that the job of accountants is to measure and keep track of explicit costs and explicit legal claims as best they can; to introduce cost changes that do not reflect an actual transaction might put all accounting on an estimate, guess, and valuation basis. Because a large part of the depreciation is taken "up front," future changes in the value of the asset are of much less importance.

Other Assets

Tangible assets are physical and measurable assets that are used in a company's operations. Current assets and fixed assets like property, plant, and equipment, are tangible assets. Intangible assets could include internet domain names, performance events, licensing agreements, service contracts, computer software, blueprints, manuscripts, medical records, permits, and trade secrets. Intangible assets add to a company's possible future worth and can be much more valuable than its tangible assets. Brand equity is an intangible asset because the value of a brand is not a physical asset and is ultimately determined by consumers' perception of the brand. For example, a consumer might be willing to pay more for a tube of Crest toothpaste rather than purchasing the store brand's toothpaste, despite it being cheaper. There can be a significant industry effect to intangible assets.

1. Technology companies, particularly within the area of computer companies, copyrights, patents, critical employees, and research and development are key intangible assets.
2. Entertainment and media companies have intangible assets such as publishing rights and essential talent personnel.
3. Consumer products and services companies have intangibles like patents of formulas and recipes, along with brand name recognition, are essential intangible assets in highly competitive markets. Coca-Cola Company is an example of an intangible asset. The value of its highly recognized brand name is virtually inestimable and is a critical driver in the Coca-Cola Company's success and earnings.
4. The health care industry tends to have a high proportion of intangible assets, including brand names, valuable employees, and research and development of medicines and methods of care.
5. The automobile industry also relies heavily on intangible assets, primarily patented technologies and brand names. For example, brand names like "Jaguar" and "Ferrari" are worth billions.

Liabilities and Stockholder Equity

The liabilities and stockholder equity section of the balance sheet shows the claims of owners and creditors against the asset values of the business. It presents the various sources from which the firm obtained the funds to purchase its assets and thereby conduct its business. The liabilities represent the claims of people who have lent money or extended credit to the firm; the ownership, capital, net worth, or equity accounts (these terms are interchangeable) represent the investment of the owners in the business. Liabilities are divided into senior debts and junior debts. The senior debts are usually secured with assets while junior debts are not secured obligations. In case of bankruptcy and liquidation, whatever debt is most senior is paid first, and whatever debt is junior is paid second. Equity holders would receive whatever is left over after the debts and other obligations have been paid.

The credit side of the balance sheet is often called the financial section or the firm's financial structure. It is especially important to the reader of finance. Many of the items found here will be discussed briefly, and they will be used in other parts of this book.

Current Liabilities

The current liabilities are those liabilities, claims, or debts that fall due within the year. Among the more common current liabilities are accounts payable, representing creditors' claims for goods or services, and notes payable or trade acceptances payable, arising out of similar economic transactions. Owed to bankers, obligations to bankers, notes payable to bank, bank loans payable or similar accounts show the amounts owing to banks for money borrowed. Usually, these arise from short-term loans, but the amounts due within the year on installment or term loans are also a current liability. Similarly, any portion of the long-term debt, i.e., bonds, mortgages, etc., maturing during the year is also carried in the current liability section. Firms that cannot make their current liability payments when they are due incur higher costs and penalties, and might have to sell supplies, inventory, and other assets at discounts or sell short term notes or borrow to raise sufficient funds. Management of the cash inflows and the outflows is an important part of the running a business in a sustainable way. Traditionally, management sought to maintain two times the current assets as the current liabilities of the firm, to maintain adequate (liquid) protection.

Accruals, a common group of current liabilities, represent claims that have built up but are not yet due, such as accrued wages, interest payable, and accrued taxes.[22] An item that bulks large for many corporations today is the amount owing on the federal and state corporation profits taxes. It appears as accrued income tax, provision for federal income tax, or other similar title.

Dividends on the common or preferred stock that have been declared but have not yet been paid are carried among the current liabilities as dividends payable.

The relationship of current liabilities to current assets is useful in many types of financial analyses and is especially important in analyzing the short-run credit position of the firm. Thus, the current liabilities are divided into the current assets to obtain the current ratio, and the current liabilities are subtracted from current assets to obtain the firm's net working capital. The larger the current ratio and the larger the net working capital relative to its total operations, the greater is the comparative safety of the firm's short-run financial position. The methods commonly used to judge the safety of the current liability coverage or the adequacy of net working capital vary with the type of firm and industry and with the judgment and analytical ability of the analyst.

Long-Term Debt

Under the classification of long-term debt, fixed liabilities, or funded debt is placed the amount the corporation owes on bond issues, mortgage notes, debentures, borrowings from insurance companies, or term loans from banks. The company may have obtained funds to acquire assets and invest in the business from these sources, and this section of the balance sheet shows the amounts still owing.

There are generally three distinctions between the long-term debt and the current liabilities. First, the items making up the long-term debt are usually more "formal" than those in the current liability section. A written legal contract or indenture describes the obligation, contains provisions for repayment under different circumstances, and details various devices for protecting the creditors against default and contains other clauses or provisions, which might work to the benefit of the debtor company. The long-term debt is also often composed of "securities," or printed certificates issued by the corporation standing for evidence of the ownership of the debt which can be freely traded or negotiated. The second important distinction is that the long-term debt will not mature for at least a year and usually for some time longer than that. Moreover, the current liabilities are generally composed of recurring items, whereas the long-term obligations are incurred only occasionally. Third, the majority of long-term obligations carry some interest charge, whereas most current liabilities do not.

Deferred Credits

Somewhere between the liability and equity section of the balance sheet we often find a category titled deferred credits or perhaps deferred, prepaid, or unearned income. These show a source of funds or assets for which the firm has not as yet performed any service. For example, suppose a company received a cash prepayment for a job on which work is not yet completed. The deferred credit classification does not mean that the firm owes money for this payment but that it owes completion of the project. Furthermore, if the firm has made this contract on a normal basis, some part of the prepayment will not be covered by services or goods, but will revert to the firm as profit. As the contract progresses, the accountants will normally analyze the results to date and apply a proportionate part of the prepayment to expenses, another part to profits (if any), and last leave (among the deferred credits) only that proportion which represents the uncompleted part of the contract.

Among the deferred credits are any rent payments made in advance by the company's tenants. Quite often an unamortized bond premium can appear. If the company at one time issued bonds bearing a coupon interest rate above that of the going market rate at the time, the bonds would have sold on the market at some premium above their face or stated value. This premium is in a sense, then, compensation to the company for their generous interest rate. The premium is amortized over time as reduction against the stated interest payment, i.e., it becomes an offset against an expense item. The unamortized portion is carried as a deferred credit since it is a sort of partial prepayment of the company's obligation to pay interest. Provisions for risks and charges and deferred taxes also are in total liabilities.

Common Equity

Common stock, capital surplus, revaluation reserves, retained earnings, ESOP guarantees constitute the majority of common equity.[23] Those terms and other variants are used interchangeably; they mean approximately the same thing, and you should learn to identify these terms so that you will not be confused if one or the other is used. This section of the balance sheet contains the items making up the ownership claims against the business. It represents the original investments of the owners plus any earnings they have retained in the business, or less any accumulated losses the business might have suffered. Retained earnings change every year by the net income of the firm, less the dividend payment. The firm's retained earnings are the primary source of the firm's cash flow and growth. Common equity is often referred to as the capital section of the balance sheet (Schwartz (1962)).

Preferred Stock

The amount shown as preferred stock represents the par or stated value of the various types of preferred stock issued, sold, and outstanding. The class of preferred stock is usually identified by its stated yearly dividends, i.e., the $7 p'f'd., the $5 p'f'd., or perhaps in percentage terms the 4 1/2 percent or 5 percent. Although the creditors can classify or consider the preferred issues simply as another form of equity, these shares have a prior claim on dividends and usually in case of dissolution have a claim prior to the common stockholders on the assets. They therefore, from the viewpoint of the common shareholders, take on some of the aspects of a creditor claim.

Common Stock

The common stock account shows the par value or stated value of the common stock issued, sold, or outstanding. The capital stock account often represents more than the "original" investment, because common stock might have been issued and sold periodically on the primary market as the firm raised funds to expand and to improve its equity base. In another sense the capital stock account might represent less than the original investment, for if in time the value of the firm's stock went over par, or over its stated value, and new issues were sold at a higher price, the difference is classified as capital or paid-in surplus. However, the new purchasing stockholders, at least, might well consider this amount part of their original investment.

Capital Surplus

The capital surplus accounts, paid-in surplus, donated surplus, premium on capital stock, or perhaps investment in excess of par value of capital stock represent funds or assets given to the business on behalf of the ownership interests. These funds or assets do not arise out of the "normal" operations, of the firm but out of certain financial transactions. For example, a capital surplus would arise if someone, perhaps but not necessarily a stockholder, were to donate the firm some assets without asking for stock or other legal obligation in return. Most commonly, capital surplus arises when a firm floats an issue of its stock at more than par value. After a company has been in operation for a time, the market value of the stock is more likely than not higher than the par value. The amount the company obtains in excess of the par value is classified as paid in surplus or premium on stock issued, etc. A premium on the issue price of either preferred or common stock is considered capital surplus.

Retained Earnings

Retained earnings or earnings surplus shows the amount that the firm has reinvested in the business out of earnings that could otherwise have been paid out in common stock dividends. These surpluses differ from capital surplus in that they arise out of accumulated retained earnings and not out of financial transactions.[24] If the firm's operations over time show accumulated losses rather than earnings, there is, of course, no earned surplus account but an accumulated deficit, which is subtracted from the other capital accounts on balance sheet. (One year's unsuccessful operation might not create a deficit on the balance sheet, since the losses of the current period might be more than covered by previously accumulated surplus.) The earned surplus accounts are basically derived from this equation: earnings minus losses minus dividends equal earned surplus. For the account to be negative, accumulated losses and dividends over time have to exceed the amounts earned.[25] The retained earnings-to-total assets ratio is a component of the Altman Z bankruptcy prediction model shown later in this chapter.

Classified among the earned surplus accounts are the so-called appropriated surplus accounts, or surplus reserves. These accounts bear such titles as reserve for contingencies, reserve for plant expansion, or reserve for sinking fund. These reserves are set up on the books by debiting earned surplus and crediting the reserve; i.e., the unappropriated earned surplus account is reduced, and a surplus reserve account is created by an equal amount. The purpose of these accounts is to warn the reader of the balance sheet that certain events are likely to occur that will reduce either the liquidity of the firm or both the liquidity and the asset holdings of the firm. For example, if the firm has embarked on a program of plant expansion and plans to use some of its present cash or marketable securities to finance the expansion, then the completion of the building program will reduce liquidity but not total assets. If, however, the firm faces a contract renegotiation with the possibility that it might have to relinquish some of its funds, the firm might lose cash without obtaining any other asset to offset it. If the management of the firm has plans for or is worried about losing some of its liquid assets, obviously it cannot distribute them as cash dividends. This is what the appropriated surplus accounts are supposed to show.

The question arises whether the existence of the appropriated surplus accounts means that the rest of the surplus – the unappropriated earned surplus – is available, for dividends. Unfortunately, the answer is no. The unappropriated earned surplus could have accumulated over the years, and the funds that it represents might have long since gone into physical plant, inventory, or other operating assets. Although legally possible, it might not be financially possible to payout any large percentage of the "free" surplus without crippling the operations of the company. As long as the company maintains the level of its operations, for all practical purposes the earned surplus can be as permanently committed to the business as the rest of the capital accounts. In fact, cash is the source of dividends or stock buy backs, not the surplus account.

2.5 Why Issue Debt? Calculating the Return on Equity

Leverage is profitable if the rate of earnings on total assets is higher than the going rate of interest on the debt. Of course, the risk to the stockholders of loss and failure in case of a downturn must always be considered. It is generally felt that to finance safely with leverage, the stability of the earnings, or better the cash flow, is more important than its level. The reader is invited to follow the Lerner-Carleton derivation of a return on equity and the issue of leverage. The operating return on assets, ROA, or R is the ratio of the firm's EBIT to total assets. The firm pays interest on its liabilities, L, with a coupon rate of r.

EBIT = R (Total Assets)

Operating Income = EBIT = R (Liabilities + Equity) = R(L+E)

Less Interest Paid = -I = r(Liabilities) = rL

Earnings before Taxes = EBT = R(L+E) – rL

Taxes Paid = -Taxes = t[R(L+E)-rL]

Earnings after Taxes = EAT = (1-t)[R(L+E) – rL]

The return on equity is given by earnings after taxes divided by equity, and is a positive function of the liabilities-to-equity ratio.

$$ROE = \frac{EAT}{E} = \frac{(1-t)\,[R(L+E) - rL]}{E}$$

$$= \frac{(1-t)\ \ [RL + RE - rL]}{E}$$

$$= (1 - t)[R + (R - r)\frac{L}{E}] \tag{1}$$

Thus, as long as the return on asset exceeds the cost of debt, then the return on equity rises linearly with leverage. Leverage is extremely important to the firm's stockholders. The choice of capital structure must be made with management's perception of the return on assets and its expected cost of debt. [26] If r is the interest rate on long-term debt, B, and the taxes paid on corporate earnings, t_c, stock income t_{ps}, and debt, t_{pd}, then the value of the firm rises with leverage, because of the interest deductibility of taxes.

To see why this is so, we know that the total income available to both stockholders and debt holders is

$$\left[(EBIT - rB)(1 - t_c)(1 - t_{ps})\right] + \left[rB(1 - t_{pd})\right]$$

where the after-tax income to stockholders is depicted by the first major bracketed term and that to the debt holders by the second, and EBIT is earning before interest and taxes. Rearranging, we obtain

$$(EBIT - rB)(1 - t_c)(1 - t_{ps}) + rB(1 - t_{pd})\left[1 - \frac{(1 - t_c)(1 - t_{ps})}{1 - t_{pd}}\right]$$

The first part of this equation is the income from an unlevered company after corporate and personal taxes have been paid. If an individual buys a bond for B, he or she receives, in the construct of our previous examples, $rB(1 - t_{pd})$ annually forever. Therefore, the value of the last part of the second equation is

$$\left[1 - \frac{(1 - t_c)(1 - t_{ps})}{1 - t_{pd}}\right] B \tag{2}$$

The deductibility of interest for tax purposes enhances the value of the firm.

Book Value of Common Stock

The book value of the common stock is derived from the balance sheet data. The book value of stockholder wealth is divided by the number of shares of stock outstanding. The market value of the stock is equal to the stock price per share times the number of shares outstanding. Alternatively, the book value per share equals the stated or par value of the common shares issued and outstanding, plus all the capital surplus, earned surplus, and surplus reserve accounts, less any liquidation premium or accrued dividends on the preferred shares, divided by the number of common shares outstanding. The book value of common stock did not warrant much analysis in the original Graham and Dodd (1934), but is a major component in the stock valuation analysis of Fama and French (1992) and Chen, Lakonishok and Hamao (1991). The ratio of the market value of stock divided by the book value of the stock, along with the stock beta, which is estimated in Chapter 3, and the market capitalization of the firm (size) are three determinants of stock returns in Fama and French (1992).

Preferred stock is not included in book value either as a sum or as part of the divisor. If the term book value is used, it is usually understood as referring to the book value of the common stock, since the concept of book value of the preferred is not important or useful. Except in rare instances the preferred stockholders are not conceived of as having any ownership or interest in the surplus accounts; it is the common shares pro rata equity in the surplus account that lends meaning to the concept of book value.

If a company owns a minor part of another firm, it can be represented by a Minority Interest item. The book value of the subsidiary company's minority stock (i.e., those shares of stock the patent company does not own) will be placed on the balance sheet midway between the liability and capital sections, since this account, usually entitled interest of minority shareholders in consolidated subsidiaries, is somewhat of a hybrid.

The IBM balance sheets from the 1997–2016 time period are shown in Table 2.3. Total assets grew from $81.5 billion in 1997 to $117.47 billion in 2016. Current assets grew only from $40.4 billion in 1997 to 43.9 billion in 2016. Current assets were 49.6 percent of total assets in 1997 and 37.4 percent of total assets in 2016. Long-term assets, intangible assets, have become the dominant asset at IBM. Long-Term Debt and Provisions for Risks and Charges, $34.66 and $22.98 billion, respectively, in 2016, have become the primary liabilities of IBM, amounting to 49.6 percent of total liabilities and shareholder equity in 2016. Long-Term Debt and Provisions for Risks and Charges is primarily underfunded pension liabilities, $6.8 billion in the US and $17.6 billion in non-US liabilities.

IBM's Provisions for Risks and Charges is growing at approximately the same rate as its long-term debt. Pension liabilities are very different for IBM in 2017 than in 1997. Shareholder equity, at $18.25 billion, was only 15.6 percent of total liabilities and shareholder equity in 2016. IBM, a highly leveraged firm in 2007, is even more highly leveraged presently. The use of leverage at IBM inflates the Return on Equity, as we will shortly discuss.

Table 2.3: IBM Balance Sheets

	DEC '16	DEC '12	DEC '08	DEC '04	DEC '00	DEC '97
			Restate	Restate		
All figures in billions of US Dollar except per share items						
Balance Sheet						
Assets						
Cash & Short-Term Investments	8.53	11.13	13.68	10.57	3.72	7.55
Short-Term Receivables	29.25	30.58	27.56	28.14	30.73	23.83
Inventories	1.55	2.29	2.70	3.32	4.77	5.14
Other Current Assets	4.56	5.44	5.07	5.12	4.67	3.90
Total Current Assets	**43.89**	**49.43**	**49.00**	**47.14**	**43.88**	**40.42**
Net Property, Plant & Equipment	10.83	14.00	14.30	15.18	16.71	18.35
Total Investments and Advances	0.70	0.93	1.85	0.97	9.31	9.03
Long-Term Note Receivable	9.44	13.02	11.42	10.95	5.71	3.72
Intangible Assets	40.89	33.03	21.10	10.23	1.63	1.77

	DEC '16	DEC '12	DEC '08	DEC '04	DEC '00	DEC '97
Deferred Tax Assets	5.22	3.97	7.27	4.67	2.97	--
Other Assets	6.51	4.83	4.58	21.87	8.14	8.22
Total Assets	**117.47**	**119.21**	**109.52**	**111.00**	**88.35**	**81.50**
Liabilities & Shareholders' Equity						
ST Debt & Curr. Portion LT Debt	7.51	9.18	11.24	8.10	10.21	13.23
Accounts Payable	6.21	7.95	7.01	9.44	8.19	5.22
Income Tax Payable	3.24	--	2.74	--	4.83	2.38
Other Current Liabilities	19.32	26.49	21.44	22.24	13.18	12.68
Total Current Liabilities	**36.28**	**43.63**	**42.44**	**39.79**	**36.41**	**33.51**
Long-Term Debt	34.66	24.09	22.69	14.83	18.37	13.70
Provision for Risks & Charges	22.98	25.25	23.01	18.22	7.71	6.80
Deferred Tax Liabilities	0.42	0.45	0.27	1.77	1.62	1.49
Other Liabilities	4.75	6.82	7.54	4.71	3.61	6.19
Total Liabilities	**99.08**	**100.23**	**95.94**	**79.32**	**67.73**	**61.68**
Preferred Stock (Carrying Value)	0.00	0.00	0.00	0.00	0.25	0.25
Common Equity	18.25	18.86	13.47	31.69	20.38	19.56
Total Shareholders' Equity	**18.25**	**18.86**	**13.47**	**31.69**	**20.62**	**19.82**
Accumulated Minority Interest	0.15	0.12	0.12	0.00	0.00	0.00
Total Equity	**18.39**	**18.98**	**13.59**	**31.69**	**20.62**	**19.82**
Total Liabilities & Shareholders' Equity	**117.47**	**119.21**	**109.52**	**111.00**	**88.35**	**81.50**
Per Share						
Book Value per Share	19.29	16.88	10.06	19.26	11.56	10.10

Source: FactSet Fundamentals

The corporate balance sheet allows the reader and investor to view the firm at a particular point in time. The reader sees the composition of the firm's capital structure, as well as the composition of current and fixed assets. Management's decisions on capital investment, dividend policy, and reliance on external financing can alter the presentation of corporate balance sheets and several financial ratios that many investors calculate to assess the firm's financial health over time.

The balance sheet of a company and the income statement are related. The balance sheet is an accounting snapshot at a point of time. The income, profit and loss, or operating statement is a condensation of the firm's operating experiences over a given period of time.[27] It depicts certain changes that have occurred between the last balance sheet and the present one. The balance sheet (position statement) and the income statement can be reconciled through the earned surplus account. If this reconciliation is presented formally, it becomes the surplus statement.

The income statement is very important. The balance sheet depicts how much assets historically have been invested in a firm; the operating statement indicates how successful (whether by efficiency, daring, or chance) the company has been in making a return on the assets committed to it.

Retained Earnings vs. Dividends

Through 2017, corporate profits in the year earned were taxed by the federal government at a rate going up to 35 percent, but the "Tax Cuts and Jobs Act" reduced the corporate tax rate to a flat 21 percent while individuals and married couples continue to pay up to 37 percent. Any distribution of dividends is taxed as additional income to the recipient at the appropriate personal income tax bracket rate.[28] This is what constitutes the "double taxation of corporation income."

Although the disparity between the corporation profits rate and the personal income tax rate, and the fact that they are applied separately, might be economically detrimental to some holders of corporation shares, it might benefit others. For individuals in very high income brackets, the 21 percent corporation profits tax might be lower than their personal tax rate. If the corporation retains some profits and pays dividends out of these at a later date – for example, at retirement, when the individual's tax bracket might have fallen – there can be a net savings of overall tax. More usual is the use of retained earnings to generate more earnings (and potentially higher dividends) so that, for example, the stock can be sold at a capital gain equivalent to the retained earnings.

2.6 Annual Cash Flow Statement

The corporate *income* statement allows an investor to view the firm's operating activities during the past year (or quarter). The investor sees the composition of the firm's revenues, operating costs, and the components and determination of net income. The income statement provides a picture of the firm's operations during the past year. The "bottom line" of the income statement is the firm's net income or after-tax profits.

In contrast, the annual *cash flow* statement allows the investor to see how the firm generates and uses its cash flow. The firm's sources of cash flow from its operations are positive net income, having depreciation and other non-cash expenses, net decreases in its current assets and net increases in its current liabilities (funds from operating activities), decreases in its long-term (fixed) assets (funds from investing), and issuing new debt or equity (funds from financing). The firm uses its cash flow to pay dividends, engage in capital expenditures, repurchase debt and equity, decrease its current liabilities, or increase its current and / or fixed assets. The firm's sources of funds must equal its uses of funds.[29]

Stockholders prefer to see a firm's cash flow derived from profits, not depreciation, because depreciation is an expense that serves to provide the firm with cash flow to replenish its capital investment. Please note that in the year 2016, IBM produced $11.872 billion of net income, which was used to engage in capital expenditures of $4.150 billion and pay dividends of $5.256 billion. IBM also had depreciation of $4.381 billion in 2016. In Table 2.4, you can see that IBM issued $2.763 billion of net new debt, and repurchased $3.502 billion of stock. The use of debt provides a higher return to stockholders than using equity, when the firm's return on total assets exceeds the cost of debt.

Table 2.4: IBM Cash Flow Statements

	DEC '16	DEC '12	DEC '08	DEC '04	DEC '00	DEC '97
All figures in millions of US Dollar except per share items						
Cash Flow						
Operating Activities						
Net Income / Starting Line	11,872	16,604	12,334	8,448	8,093	6,093
Depreciation, Depletion & Amortization	4,381	4,676	5,450	4,915	4,995	5,001
Deferred Taxes & Investment Tax Credit	-1,132	797	1,900	2,081	29	358
Other Funds	606	-41	321	-483	-792	-718
Funds from Operations	15,727	22,036	20,005	14,961	12,325	10,734
Changes in Working Capital	1,231	-2,450	-1,193	362	-3,051	-1,869

	DEC '16	DEC '12	DEC '08	DEC '04	DEC '00	DEC '97
Net Operating Cash Flow	**16,958**	**19,586**	**18,812**	**15,323**	**9,274**	**8,865**
Investing Activities						
Capital Expenditures	-4,150	-4,717	-4,887	-5,056	-6,181	-7,107
Net Assets from Acquisitions	-5,679	-3,722	-6,313	-1,738	0	0
Sale of Fixed Assets & Businesses	-30	1,009	421	1,336	1,619	1,130
Purchase/Sale of Investments	-225	-967	1,510	112	314	-178
Other Funds	-892	-607	-16	0	0	0
Net Investing Cash Flow	**10,976**	**-9,004**	**-9,285**	**-5,346**	**-4,248**	**-6,155**
Financing Activities						
Cash Dividends Paid	-5,256	-3,773	-2,585	-1,174	-929	-783
Change in Capital Stock	-3,502	-11,995	-10,578	-5,418	-6,073	-6,251
Issuance/Reduction of Debt, Net	2,763	2,252	-2,444	-1,027	643	3,944
Other Funds	204	1,540	3,773	0	0	0
Net Financing Cash Flow	**-5,791**	**-11,976**	**-11,834**	**-7,619**	**-6,359**	**-3,090**

Source: FactSet Fundamentals

A firm can find itself in trouble with "excessive debt" when its return on assets falls below its cost of debt. IBM new debt issues have been positive in the 2000–2016 period (see Table 2.5), whereas IBM has not issued stock. IBM has re-purchased more stocks than it has issued during the 1997–2016 period. [30]

Table 2.5: Corporate Ratios for IBM and Dupont for Selected Years

	5 Yr AVG	DEC '16	DEC '12	DEC '08	DEC '04	DEC '00	DEC '97
Profitability (%)							
Gross Margin	48.08	47.03	47.43	43.46	36.89	36.68	38.99
SG&A to Sales	29.22	31.61	27.35	27.39	24.82	23.52	27.40
Operating Margin	18.86	15.42	20.08	16.06	12.07	13.16	11.59
Pretax Margin	19.41	15.45	20.97	16.13	12.49	13.05	11.50
Net Margin	16.12	14.87	15.89	11.90	8.77	9.16	7.76
Free Cash Flow Margin	15.21	16.76	14.84	14.13	11.38	4.14	2.64
Return on Assets	12.52	10.42	14.09	10.73	7.91	9.20	7.49
Return on Equity	86.11	73.10	85.15	58.82	29.33	39.35	29.40
Return on Common Equity	86.11	73.10	85.15	58.82	29.33	39.73	29.67
Return on Total Capital	26.79	20.74	32.06	22.20	16.22	16.51	13.36
Return on Invested Capital	30.93	23.62	38.64	28.14	18.89	21.98	18.74
Valuation (x)							
Price/Sales	1.92	1.99	2.12	1.12	1.75	1.74	1.35
Price/Earnings	11.83	13.40	13.33	9.42	20.00	19.14	17.41
Price/Book Value	10.27	8.60	11.35	8.37	5.45	7.35	5.18
Price/Tangible Book Value	--	--	--	--	8.31	7.99	5.69
Price/Cash Flow	10.02	9.38	11.30	6.18	10.99	16.61	11.93

	5 Yr AVG	DEC '16	DEC '12	DEC '08	DEC '04	DEC '00	DEC '97
Price/Free Cash Flow	12.70	11.88	14.28	7.94	15.38	42.11	51.05
Dividend Yield (%)	2.66	3.31	1.72	2.26	0.71	0.60	0.74
Enterprise Value/EBIT	11.84	15.48	11.26	8.03	15.02	15.04	13.29
Enterprise Value/EBITDA	9.35	11.42	9.21	6.05	10.55	10.52	8.58
Per Share							
Sales per Share	87.86	83.36	90.45	75.00	56.35	48.78	38.83
EBIT (Operating Income) per Share	16.63	12.85	18.17	12.05	6.80	6.42	4.50
EPS (recurring)	15.06	13.17	14.83	9.27	5.21	4.44	3.00
EPS (diluted)	14.36	12.38	14.37	8.93	4.93	4.44	3.01
Dividends per Share	4.35	5.50	3.30	1.90	0.70	0.51	0.39
Dividend Payout Ratio (%)	30.94	44.42	22.96	21.28	14.20	11.49	12.90
Book Value per Share	16.91	19.29	16.88	10.06	18.08	11.56	10.10
Tangible Book Value per Share	-18.45	-23.94	-12.69	-5.70	11.86	10.63	9.19
Cash Flow per Share	16.90	17.69	16.95	13.61	8.97	5.12	4.38
Diluted Shares Outstanding	**1,041.98**	**958.71**	**1,155.45**	**1,381.77**	**1,708.87**	**1,812.12**	**2,021.87**
Basic Shares Outstanding	1,035.09	955.42	1,142.51	1,359.77	1,674.96	1,763.04	1,966.57
Total Shares Outstanding	1,014.78	945.87	1,117.37	1,339.10	1,645.59	1,762.90	1,936.18
Asset Turnover Analysis (x)							
Cash & ST Investments	9.38	9.56	9.07	7.13	10.57	18.51	10.01
Receivables	3.01	2.77	3.48	3.68	3.38	3.03	3.34
Inventories	23.33	27.28	22.51	21.84	19.42	11.62	8.70
Current Assets	1.91	1.85	2.08	2.03	2.10	2.03	1.94
Fixed Assets	7.45	7.41	7.50	7.05	6.45	5.15	4.39
Total Assets	0.78	0.70	0.89	0.90	0.90	1.01	0.97
DuPont Analysis							
Asset Turnover (x)	0.78	0.70	0.89	0.90	0.90	1.01	0.97
x Pretax Margin (%)	19.41	15.45	20.97	16.13	12.49	13.05	11.50
= Pretax Return on Assets (%)	15.15	10.83	18.60	14.54	11.26	13.12	11.10
x Tax Rate Complement (1-Tax Rate)	83.85	96.36	75.82	73.79	70.24	70.17	67.50
= Return on Assets (%)	12.52	10.42	14.09	10.73	7.91	9.20	7.49
x Equity Multiplier (Assets/Equity)	6.94	7.01	6.04	5.48	3.71	4.27	3.92
= Return on Equity (%)	86.11	73.10	85.15	58.82	29.33	39.35	29.40
x Earnings Retention (1-Payout)	69.06	55.58	77.04	78.72	85.80	88.51	87.10
= Reinvestment Rate (%)	59.46	40.76	65.80	46.50	25.25	34.83	25.62
Note: EBIT Return on Assets (%)	14.71	10.81	17.81	14.48	10.88	13.23	11.19
Note: Interest as % Assets	0.50	0.66	0.48	0.61	0.13	0.81	0.89

	5 Yr AVG	DEC '16	DEC '12	DEC '08	DEC '04	DEC '00	DEC '97
Operating Efficiency (x)							
Revenue/Employee	0.21	0.19	0.22	0.24	0.26	0.24	0.29
Net Income/Employee	0.03	0.03	0.04	0.03	0.02	0.02	0.02
Assets/Employee	0.27	0.28	0.26	0.25	0.30	0.24	0.30
Receivables Turnover (x)	3.01	2.77	3.48	3.68	3.38	3.03	3.34
Inventory Turnover (x)	23.33	27.28	22.51	21.84	19.42	11.62	8.70
Payables Turnover (x)	6.65	6.92	6.63	7.78	6.83	7.66	9.45
Asset Turnover (x)	0.78	0.70	0.89	0.90	0.90	1.01	0.97
Working Capital Turnover (x)	11.35	10.50	17.99	15.78	13.43	11.83	11.36
Liquidity (%)							
Current Ratio	1.22	1.21	1.13	1.15	1.18	1.21	1.21
Quick Ratio	1.17	1.17	1.08	1.09	1.10	1.07	1.05
Cash Ratio	0.24	0.24	0.26	0.30	0.27	0.10	0.23
Cash & ST Inv/Current Assets (%)	19.98	19.43	22.51	26.34	22.50	8.48	18.69
CFO/Current Liabilities (%)	45.48	46.75	44.90	44.33	38.50	25.47	26.46
Coverage (x)							
EBIT/Interest Expense (Int. Coverage)	30.18	15.82	35.76	24.19	81.31	15.79	11.97
EBITDA/Interest Expense	38.50	21.50	45.11	32.83	119.01	23.19	19.37
Fixed-charge Coverage Ratio	30.18	15.82	35.76	24.19	81.31	15.20	11.52
Leverage (%)							
LT Debt/Total Equity	198.34	189.93	127.72	168.50	49.85	89.08	69.12
LT Debt/Total Capital	56.89	57.36	46.21	47.88	28.15	37.34	29.30
LT Debt/Total Assets	27.17	29.50	20.21	20.72	13.58	20.79	16.81
Total Debt/Total Assets	33.22	35.90	27.91	30.97	21.00	32.34	33.04
Net Debt/Total Equity	184.42	184.37	117.39	156.09	41.54	120.51	97.76
Total Debt/Equity	241.05	231.11	176.40	251.95	77.07	138.56	135.88
Total Debt/Total Capital	69.66	69.80	63.82	71.59	43.53	58.08	57.61

Source: FactSet Fundamentals

Five-year ratio averages can be very important in financial analysis of destressed firms. Declining five-year profitability ratios are associated with bankruptcy risk, see Beaver (1966) and Barth, Beaver and Landsman (1998). In the Barth, Beaver, and Landsman (BBL, 1998) manuscript, the market value of equity is associated with the book value of debt and net income. The BBL valuation model also measured the market value of equity as a function of the book value of equity and unrecognized net assets (such as research and development expenditures and advertising expenditures). When financial health is lower, then net income variable coefficient and incremental explanatory power is lower. When financial health is lower, then the coefficient of the book value of equity variable is lower. The BBL 1998 study examined US bankruptcies listed on the 1994 Compustat tape that occurred between 1974 and 1993. BBL reported that 77 of the 396 bankrupt firms had negative book value of equity in the year prior to bankruptcy. BBL reported that the coefficient and explanatory power of on the net income variable decreased in the five years prior to bankruptcy. Pharmaceutical firms, because of their high levels of research and development expenditures with large intangible assets, are priced more on earnings than book value of equity. In a later study, Barth, Correia, and McNichols (BCM, 2012) reported that the predictive power of financial ratios had declined during the 1962–2002 time period for over 1850 bankrupt firms. Firms with losses are still ten times more likely to declare bankruptcy than firms with positive net income. Firms with higher profitability, less leverage, and higher EBITDA to total liabilities are associated with lower bankruptcy probabilities. The BCM study noted that the presence of intangible assets, measured by research and development intensity, had a highly significant impact on bankruptcy prediction. Higher R&D expenditures led to lower predictive power for the accounting-based model.

Why do firms use current liabilities and long-term debt rather than issue common stock to finance growth? A rapidly growing and highly profitable firm can grow net income by earning more on its return on investments, referred to as the reinvestment rate, than the interest the firm pays on its current liabilities and long-term debt. If the reinvestment rate exceeds the interest rate, then the return on equity, ROE, increases as it increases its leverage (proportionally), until the dangers of bankruptcy increase.

TDTA Ratio

Total debt to total assets, TDTA, is calculated by the sum of the firm's current liabilities and long-term debt, divided by its total assets. If the reinvestment rate, R, exceeds the cost of debt, r, interest expenses divided by total assets, then the use of leverage increases the return on equity. In plain English, the use of leverage, or "other people's money' (OPM), is great as long as you can pay the interest charges and repay the debt at its maturity. The excess of the reinvestment rate of assets relative to the cost of debt represents profits to the owners of the firm, its stockholders. If a firm cannot pay its interest, then the firm is in default and might be forced to declare bankruptcy.

In Table 2.6, we calculate the current analysis ratios and the general analysis ratios for IBM, Dominion Energy (formerly known as Dominion Resources), and Facebook for selected years during the 1963–2014 period, using the Wharton Research Data Services (WRDS) Compustat database. Mature and profitable firms may have TDTA ratios that are higher than less mature firms, such as Facebook.

Table 2.6: Corporate Debt and Equity Ratios

Year	Company	CATA	CLTA	TDTA	SEQTA
1963	IBM	0.478	0.150	0.330	0.671
1963	Dominion	0.481	0.048	0.520	0.428
1970	IBM	0.397	0.220	0.287	0.696
1970	Dominion	0.040	0.060	0.569	0.402
1980	IBM	0.372	0.244	0.323	0.616
1980	Dominion	0.085	0.128	0.465	0.386
1985	IBM	0.495	0.217	0.293	0.608
1985	Dominion	0.085	0.128	0.465	0.386
1990	IBM	0.444	0.286	0.425	0.489
1990	Dominion	0.071	0.085	0.485	0.399
1995	IBM	0.507	0.394	0.520	0.280
1995	Dominion	0.079	0.100	0.442	0.391
2000	IBM	0.497	0.412	0.620	0.233
2000	Dominion	0.200	0.259	0.616	0.256
2010	IBM	0.424	0.376	0.550	0.203
2010	Dominion	0.126	0.134	0.503	0.286
2010	Facebook	0.751	0.130	0.253	0.723
2012	IBM	0.415	0.366	0.568	0.158
2012	Dominion	0.110	0.166	0.525	0.231
2012	Facebook	0.746	0.070	0.201	0.778
2014	IBM	0.425	0.336	0.635	0.101

Year	Company	CATA	CLTA	TDTA	SEQTA
2014	Dominion	0.103	0.132	0.534	0.213
2014	Facebook	0.340	0.035	0.038	0.898

Source: WRDS Database: IBM, Dominion, Facebook, Selected Years

Sources of Net Working Capital

Since net working capital is the current assets minus the current debt, net working capital (by definition) must be supplied by equity and/or long-term debt. Thus, net working capital is financed from permanent or semi-permanent sources.

Working capital is not a static item, however, but expands and contracts with the level of the firm's operations. (Gross working capital is likely to show much wider proportionate changes.) Only occasionally does an established firm go out to the capital markets for additional working capital; normally, working capital is increased or depleted by internal operations of the business.

The main financial items affecting the level of working capital fall within two categories. Those that *increase* net working capital funds are:

1. net operating profits
2. current depreciation charges insofar as they are covered by revenues over variable costs
3. sales of fixed or other assets for current funds
4. issuing new bonds or stocks if the funds obtained are not all used to acquire fixed assets

Financial events that *decrease* net working capital funds are:

1. operating losses
2. payment of dividends
3. purchase of fixed or other noncurrent assets
4. repayment of long-term debt
5. buybacks, i.e. repurchasing the corporation's own shares with available funds

2.7 Ratio Analysis and Working Capital

Ratio analysis is an alternative to the flow of funds method of working capital analysis, although the two can be used to supplement each other. Ratio analysis is the older and possibly the more popular approach than the flow of funds method of management, and it is the most readily available approach for credit managers of other companies, or other outsiders. A person within the firm sometimes finds other analytical tools more useful. Ratio analysis consists of studying ratio or percentage relationships of meaningful financial data. The results are compared (1) with standard ratios – i.e., the averages of similar firms, (2) with the firm's ratios in previous years, or (3) with some implicit standards existing in the mind of the analyst. In the hands of a skilled practitioner, both external analysis (comparisons to standard ratios) and internal analysis (i.e., trends and relationships of the ratios within the company) can be revealing.

Guerard and Schwartz (2007) presented, calculated, and discussed thirteen ratios that were the most generally used, as shown in Table 2.7. The first six were most relevant for current analysis. The remaining seven reveal more general relationships. These ratios, and many more, are routinely available from FactSet Research Systems for institutional investors and money managers.

Table 2.7: Commonly Used Ratios

Name	Abbreviation
Current Analysis Ratios	
Current Ratio	CR
Acid Test	AT
Sales/Inventory	SI

Name	Abbreviation
Sales/Receivables	SR
Sales/Net Working Capital	SNWK
General Analysis Ratios	
Financial Structures Ratios	
Total Debt/Assets	TDA
Operating Ratios	
Sales/Total Assets	SA
Net Profit/Total Assets	ROA
Net Profit/Tangible Net Worth	ROTNW
Net Profit/Equity	ROE
EBIT /Interest	TIE
Composite Firm Relative Valuation Ratios	
DuPont Analysis	DuPontA
Altman – Z Model	NewZ

Current Analysis Ratios

Current Ratio

The current ratio is obtained by dividing the current liabilities into the current assets. It indicates how many times current liabilities are covered by gross working capital. The higher the current ratio, the more conservative the current financial position of the firm.

A two-to-one ratio is a rule-of-thumb benchmark indicating a minimum level of the working capital position. Other circumstances must always be considered; no financial analysis can proceed rigidly. A ratio below two does not necessarily make the firm unsafe, nor does a current ratio well over two ensure financial soundness. Much depends on the collectability and time structures of the firm's receivables and the type and quantity of inventory the firm carries. Public companies often have a current ratio of one to one or below.[31] In an electric utility company, for example, the low current ratio is possible because of its minimum inventory requirements and the stability of its revenues and cash flow.

Acid Test

The acid test, or quick ratio, is obtained by dividing current liabilities into the firm's net receivables and cash. This ratio highlights the firm's short-term liquidity position. The rule-of-thumb measure of a satisfactory acid test ratio is one to one. From an obverse point of view, the acid test ratio tends to indicate the amount of inventory in the working capital position of the firm. For example, if the current ratio is 3 to 1 and the acid test is only .85 to 1, the inventory account probably constitutes a heavy proportion of the current assets.

Sales to Inventory: Approximate Inventory Turnover

This ratio is obtained by dividing the inventory into the sales figure. The result is useful in analyzing how rapidly the firm's inventory is sold. A slow turnover – relative to the type of business or its own previous performance – might indicate that the firm is overstocked, or that the inventory contains too many old or out-of-style items, or that the management is speculating in inventory. Again, as in the case of the receivable turnover, the inventory turnover figure can be divided into 360 days to get an average of how many days it takes for a given dollar amount of merchandise to be turned into an equivalent amount of sales.

Many analysts prefer to reserve the term "inventory turnover" for the ratio of inventory divided into the cost of goods sold. The ratio then indicates the true physical turnover of the inventory. Since the sales figure contains the gross markup (profit) over cost, the sales over inventory ratio overstates the actual physical turnover of the goods. The higher the customary gross margin, or markup over cost in the sales figure, the better the inventory into sales ratio appears in comparison to the true turnover ratio.[32]

Unfortunately, the figure for the cost of goods sold is not always as available as the amount of sales. Thus, the standard ratios are more often based on sales. Ratio analysis, in any case, is not an exact science but is based on historical or intra-industry comparisons. The ratio serves its purpose as long as it depicts a logical relationship and comparisons using it are made on a consistent basis.

Sales to Receivables

The Receivable Turnover Ratio

This ratio is obtained by dividing credit sales by the outstanding trade accounts and trade notes receivable, and indicates the collectability and current condition of the receivables. The higher the sales to receivables ratio, the more current are the receivables. A variant of this ratio is to divide the turnover rate into 360 (representing the approximate number of days in the year). The resulting figure gives the number of days that it takes to collect an average account. This figure can be compared to the usual terms (or allowable credit time) granted by the firm to ascertain whether the average account is collected in a period close to the credit terms.

The Cash Cycle: Sales Divided by the Receivables + Inventory

This ratio is obtained by dividing sales by the basic noncash, working assets. The ratio is usually used in its cash cycle form; the result of dividing the inventory and receivables into the annual sales are further divided into 360 days. The result is the average approximate number of days from the time material comes into the possession of the firm until it is turned into cash. If the time required by the cash cycle is shorter in comparison to that of similar firms or to the firm's own past, the firm is relatively efficient in using its current working assets. The contrary is also true. Comparing the cash cycle to the customary credit terms the firm receives from its suppliers helps determine the firm's need for net working capital. If the terms are longer than the cash cycle, the relative amount of net working capital can be minimized. If the cash cycle is equal to or longer than the customary time allowed for payment by trade creditors, the firm requires a greater amount of net working capital to stay safely in business.

Sales/Net Working Capital: Working Capital Turnover

The net working capital turnover is obtained by dividing net working capital into the annual sales. This ratio is double-edged; a high ratio can indicate either efficiency or risk. A low turnover might indicate managerial inefficiency in moving goods and collecting receivables, or it might indicate excessively conservative management – a tendency to hold redundant idle funds or a failure to use a reasonable amount of available current credit. The other edge of the ratio appears if the turnover is too high in contrast with the industry norm. It might not necessarily indicate efficiency but a tendency to take on undesirable levels of risk. An especially high net working capital turnover can indicate overtrading on current account – an attempt to carry a heavy volume of business on an inadequate current capital base. Such speculative striving on the part of the management can be dangerous to both owners and creditors.

It should be obvious that the ratios are not to be used singly but in a composite manner to fill out a financial portrait of the company. Thus, the position of the net working capital turnover can be checked against the other current operating ratios. For example, a firm with a high net working capital turnover and ordinary inventory and receivable turnovers whose current ratio is tight is likely to be overtrading. A firm with a low net working capital turnover, a normal cash cycle, and a very high acid test ratio might be holding excessive idle funds. These are only two possibilities. An experienced analyst might be able to rough out normal relationships in his head while scanning the financial figures. His instinct might lead him quickly to any items that are out of line and suggest the few ratios necessary to highlight the potential trouble spot.

2.8 General Analysis Ratios

The current analysis ratios are most important to credit managers who pass on credit sales and others who are interested in the firm's short-term position. The general analysis ratios are useful to investors, long-term creditors, and others concerned with the firm from a longer-term basis. Of course, in any case, whether the analyst's interest is short or long term, selections of pertinent ratios should be made from both groups.

Short-term creditors have sometimes loaned (given) a firm funds on the basis of a good current position, unwisely ignoring other fundamental financial analysis. For although the first grant of credit might be repaid on time, many short-term arrangements turn out to be semi-permanent as the supplier periodically renews or extends new credit to the customer firm. If fundamental financial weaknesses outside the current position are passed, they might cause failure at some later date with consequent losses to the "short-term" creditor.

The Financial Structure Ratios

These ratios, that is the percentage that each major class of financing tells of the total financing of the firm's assets. Related to the characteristics of the company's operations, these ratios indicate the overall financial risk the firm carries. They can be useful to the short- term creditor, the bondholder, or the potential investor in common stock.

Some sources give the financial ratios in the form of current liabilities to equity, long-term debt to equity, etc. This alternate form contains the same information as the financial structure, but the financial structure ratios have the advantage of presenting an immediately comprehensible picture of the composition of the credit side of the company's balance sheet.

Operating Ratios

Sales to Total Assets

The ratio of sales to total assets indicates how intensively the total assets are used in production. A high Sales-to-Total Assets ratio is an indicator of efficient use of assets. Indeed, the sales-to-assets ratio is an activity ratio, illustrating the effectiveness of using the firm's resource, see Weston and Copeland (1986). A low sales to total assets ratio in comparison with similar firms or with previous periods gives some indication of idle capacity; i.e.; excess assets compared to the level of operations.

Inter-industry comparisons of this ratio are not very useful. A wholesale distributor, for example, with no processing costs, a small margin, and a large turnover of goods shows a relatively high volume of sales to total assets. A better ratio to measure the basic concept of the rate of utilization of capital would be value-added to total asset – i.e., something approaching a capital coefficient. Unfortunately, possibly because of statistical difficulties, value-added ratios are not commonly used in financial analysis.

Net Profit to Total Assets (Preferred Ratio: EBIT to Total Assets)

This ratio is intended to relate the return of the firm to its total investment; i.e., the total assets that it has available. It has some use, but it is subject to the criticism that the relationship presented is not the most logical one and that it does not present sufficient new information. The net profit figure has already been reduced by taxes and the cost of external funds (i.e., interest); to relate this figure to total assets is an illogical relating of a net concept (net profits) with a gross concept (total assets). Moreover, the ratio does not give much independent information if the net profit on owners' equity is to be calculated too, as it usually is. Obviously the net profit rate of return (or rate of loss) on total assets is always less than that on the owners' equity. The difference depends on the relative amount of total leverage.[33]

If we are to get any clear concept of how productively the management has used the total assets, the preferable ratio is the EBIT to total assets. In addition, a comparison of this ratio to rate-of-profit on net worth indicates how well the firm has adjusted its financial mix. Comparison of these ratios (allowing for the fact that one is an after-tax return) indicates something about the profitability – but not much about the risk – of the firm's use of leverage.

Net Profit on Tangible Net Worth

This ratio is obtained by dividing the net profit by the common stock equity of the corporation.[34] It gives the rate of return made by the owners of the corporation on the book value of their investment. If this ratio is compared to the rate of return on total assets to determine whether financial leverage operated favorably for the firm, earnings after interest but before taxes (EBT) should be the numerator. Then if the earnings return on assets is properly calculated on the EBIT basis, these two ratios would be properly comparable on a financial basis. For some comparisons of profitability, therefore, it might be better to use the ratio of officers' salaries plus profits before taxes over the net worth as the basis of comparison.

The profitability ratios, that is the last three ratios just discussed, are usually more relevant for the owners and the managers of the firm than for the short-term creditors. The short-term creditors' position is safe, at least for the short run, as long as the working capital and financial ratios show the firm to be currently sound. The profit ratios, nevertheless, are important indicators of the future. If profit ratios are high (and perhaps trending higher), then perhaps some ratios do need not to be as high in comparison with other firms. Working capital deficiencies can be restored by high earnings, and debt can be met out of a high rate of cash flow. On the other hand, a series of losses can soon deplete an excellent starting working capital position. Thus, a current operating loss position that threatens to continue, or an indicated downward trend in earnings, may be taken as a danger signal, even if the existing working capital ratios seem good.

EBIT to Interest Charges (Times Interest Earned)

This ratio is obtained by dividing the firm's annual interest charges into its earnings before interest and taxes. The size of the ratio obtained indicates something of the safety of the long-term debt component of the firm's financing. The operational safety of long-term debt affects the short-term creditors and working capital management indirectly. If the coverage is good, the firm can safely operate on a relatively smaller net working capital margin. A poor or erratic coverage might cast doubt on what otherwise appears to be an adequate current position.

Financial Ratios and the Perceived Financial Health of Firms

Financial analysis often combines the information of several ratios to gain insight into a picture of the firm's health. In this chapter, we examine two composite measures of the firm's health, the "DuPont System" rate of return, dating back to the early twentieth century, and the Altman Z bankruptcy prediction model, created in 1968. The DuPont system, or measure, takes its net operating income divided by sales and multiplies by the ratio of sales-to-investment, producing a return on investment, ROI.

$$\frac{\text{NOI}}{\text{Sales}} \text{ x } \frac{\text{Sales}}{\text{Investment}} = \text{ROI} \qquad (3)$$

Stockholders should invest in firms with higher ROIs, and management could seek to maximize the DuPont ROI to maximize its stock price. The DuPont analysis uses information inherent in its return on sales and sales turnover ratios. [35]

The firm's net income is divided by total assets, sales, and equity to produce ROA, ROS, and ROE, profitability ratios, respectively. A higher ROA for the firm relative to a set of comparable firms, its industry or sector, shows that the firm's assets are used to generate higher net income and a more effective use of assets. A higher relative ROS indicates that for a given sales volume, the firm generates higher net income than comparable firms. The ROE ratio helps quantify the firm's profitability relative to its use of leverage. The very large multiple for IBM in its ROE relative to its ROA illustrates IBM's very extensive use of debt, which is driven by debt issuance and equity repurchases. The DuPont Analysis and the Altman Z Bankruptcy Model make extensive use of profitability ratios.

The multiplication of the sales turnover ratio by the ratio of earnings-to-sales produced the DuPont return on invested capital, which is still in use by the DuPont Corporation and most American firms. Total investment includes working capital, cash, inventories, and accounts receivable, and permanent investment, bonds, preferred stock, and stocks. The DuPont return on invested capital combined and consolidated financial, capital, and cost accounting. The DuPont return on total investment helped DuPont develop many modern management procedures for creating operating and capital budgeting and making short-run and long-run financial forecasts.

A second composite model is the Altman Z model, which is useful to identify potential bankrupt firms. The Altman Z score used 5 primary ratios in its initial 1968 version.

$$Z = .012\,X_1 + .014\,X_2 + .033\,X_3 + .006\,X_4 + .999\,X_5 \tag{4}$$

$$\text{Where } X_1 = \frac{\text{Current Assets} - \text{Current Liabilities}}{\text{Total Assets}}$$

$$X_2 = \frac{\text{Retained Earnings}}{\text{Total Assets}}$$

$$X_3 = \frac{\text{EBIT}}{\text{Total Assets}}$$

$$X_4 = \frac{\text{Market Value of Equity}}{\text{Book Value of Debt}}$$

$$X_5 = \frac{\text{Sales}}{\text{Total Assets}}$$

The Altman Z score used a liquidity, past profitability, (present) profitability, leverage, and the sales turnover (efficiency) ratios to produce a single score. An Altman Z score of less than 2.67 implied that the firm was not healthy. An Altman Z score exceeding 2.67 implied financial health. The Altman Z score successfully predicted impending bankruptcy for 32 of 33 firms (97 percent) in the year prior to bankruptcy, for his initial sample. The model correctly predicted 31 of 33 (94 percent) non-bankrupt firms in this sample for the year prior to bankruptcy.

Altman modified his equation in 2000 to become:

$$Z = .717X_1 + .847X_2 + 3.107X_3 + .420X_4 + .998X_5 \tag{5}$$

where X_4 is now book value of equity relative to its book value of debt. The new critical level is 2.0. The ratios of companies can be compared to all firms, companies in a sector, or broad segments of the economy, or specific industry groups.

We examine the Altman Z and its components for IBM and all companies listed on the Wharton Research Data Services (WRDS) database for selected years during the 1972–2014 time period. The median ratios for the Compustat companies are reported in Table 2.8 as well as for IBM.

Table 2.8: Corporate Ratios (Medians)

Year	Number of Companies	Current Ratio (CR)	Sales-to-Assets (SA)	Total Debt-to-Total Assets (TDA)	Return on Equity (ROE)	New Z
1972	4131	2.113	1.208	0.460	0.109	2.585
1975	6169	1.937	1.620	0.496	0.102	2.517
1980	6103	2.840	1.224	0.497	0.125	2.509
1987	7567	1.788	0.898	0.473	0.088	1.921
1990	7296	1.610	0.957	0.485	0.085	1.859
1995	9657	1.716	0.859	0.451	0.088	1.947

Year	Number of Companies	Current Ratio (CR)	Sales-to-Assets (SA)	Total Debt-to-Total Assets (TDA)	Return on Equity (ROE)	New Z
2000	9260	1.640	0.629	0.447	0.064	1.481
2008	7522	1.674	0.578	0.420	0.044	1.389
2010	7391	1.802	0.500	0.382	0.066	1.513
2014	7383	1.728	0.390	0.408	0.060	1.246
1972	IBM	2.134	0.884	0.281	0.169	3.127
1975	IBM	2.413	0.930	0.236	0.174	3.555
1980	IBM	1.521	0.982	0.323	0.217	2.936
1987	IBM	2.319	0.851	0.271	0.137	2.782
1990	IBM	1.539	0.788	0.435	0.141	2.126
1993	IBM	1.183	0.773	0.597	-0.410	1.130
1995	IBM	1.286	0.896	0.519	0.186	1.740
2000	IBM	1.205	1.005	0.620	0.392	1.852
2008	IBM	1.154	0.946	0.595	0.916	1.924
2010	IBM	1.186	0.880	0.550	0.643	2.246
2014	IBM	1.248	0.790	0.635	1.013	2.192

Source: WRDS Database: All Compustat Companies, and IBM, Selected Years

IBM has much less in current assets, cash, accounts receivable, and marketable securities than the median US company; hence, its Current Ratio (CR) is lower than the median company. A lower Current Ratio means that IBM could experience illiquidity and cash needs. IBM generated much greater sales relative to assets, SA, than the median US company post-2000; hence IBM is relatively efficient in its asset usage. IBM uses far more total debt, TDTA, than the median US firm and debt creates the opportunity to create greater returns in "good times" and far worse returns in "bad times." The greater use of debt by IBM with its profitability creates a much higher Return on Equity, ROE, than the median US firm, particularly post-2000. Finally, the Altman (New) Z identified IBM as being in great danger of financial failure in 1993, but its profitability and sales efficiency has returned IBM to a healthy financial position, much higher than the median US company. Firm ratios should be compared to comparable data of all firms in the economy, and in many, cases, to firms in its industry.

The Altman bankruptcy model has been acclaimed as one of the premier bankruptcy prediction models. The bankruptcy models often test predictions one, two, and five years ahead. The five-year averages serve much the same purpose. Firms with increasing five-year leverage ratios, decreasing five-years profitability ratios, and decreasing five-year liquidity ratios will score much lower on the Altman Z-Score and the Beaver (1966) bankruptcy.

In a recent bankruptcy prediction study, Barth, Beaver, and Landsman (BBL) estimated a valuation model in which the market value of equity was a function of the book value of equity and net income, and a second equation in which the market value of equity was a function of the book value of equity and unrecognized net assets, such as research and development (R&D) expenditures and advertising expenditures (intangible assets). BBL (1998) estimated valuation models for 396 firms that went bankrupt during the 1974–1993 time period as reported on the 1994 Compustat database. BLL reported that 77 of the 396 firms declaring bankruptcy had negative book values in the year prior to bankruptcy (p. 12).

Firms with lower financial health had lower coefficients and explanatory power associated with the net income variable than other firms. Firms with lower financial health had higher coefficients and explanatory power associated with the book value variable than other firms. Pharmaceutical firms with higher R&D and unrecognized intangible assets had significantly higher explanatory power of the net income variable rather than the book value variable. Beaver, Correia, and McNichols (BCM) reported in a larger study of over 1250 firms that went bankrupt between 1962 and 2002, that the power of the accounting-based failure model had declined, particularly for firms with higher R&D intensity. Across the book value variable, firms with low to medium book value that the highest predictive power. BCM (2012) concluded that financial statements fail to recognize changes in intangible assets or abandonment values.

2.9 Corporate Exports

Corporations seek to maximize stockholder wealth. Brealey, Myers, and Allen (2006) stated that the goal of corporate finance is to implement corporate decisions to enhance stockholder wealth. We assume that the market mechanism produces the most effective measurement of corporate strategies. Schwartz and Aronson (1964) held that the rate of return on capital and the growth rate of the economy are equal. If corporate cash flows exceeded real investment opportunities, then the firm should distribute the surplus funds in the form of dividends, interest paid on debt, buy backs, and repayment of debt. Schwartz and Aronson reported that dividends and interest paid substantially exceeded new funds raised on the capital markets. Schwartz and Aronson referred to the excess of corporate cash flows exported as "corporate exports." Guerard and Schwartz (2007) updated the analysis and confirmed the dramatic growth in corporate funds exported during the 1971–2005 period using a universe of publicly traded firms in the United States with total assets exceeding $200 million. Dividends paid and interest paid rose substantially, as did stock re-purchases.

We specifically look at dividend, debt and equity issuance and buy-back strategies, and the implied leverage decisions of firms. Fama and Babick (1968) reported that dividend policies show a high correlation of dividends with earnings, although as earnings rise, there is usually some lag before former payout levels are resumed. Buy backs are more difficult to predict. In the past, before the prevalence of buy backs, if a company had a "normal," or consistent, dividend payout policy, the prospective investor would not be far wrong if he concentrated on the trend of earnings in evaluating the worth of a share of stock. A consistent dividend policy means that the company pays a reasonably consistent ratio of dividends divided by earnings. Equity repurchases, whether privately negotiated or via a tender offer generally specify the number of shares the firm seeks to repurchase, the tender price at which it will repurchase shares, and the expiration date of the tender offer, which the firm can extend. Dann (1981) examined 143 cash tender offers to repurchase equity during the 1962–1976 period, made by 122 different firms, and reported a 22.46 percent tender offer premium, relative to the previous day of the announcement (20.85 percent relative to the one-month period before the announcement).

Large excess returns are associated with share buybacks, particularly for stocks in the 1970s. Firms repurchase equity to enhance stockholder wealth. There were marginal debt effects, and approximately 95 percent of the enhanced value accrued to stockholders. Lakonishok and Vermaelen (1990) reported excess returns to repurchases continuing through 1986, but at a (slightly) diminished rate.[36] The level of current and projected earnings impacts dividend policy and potential buy backs, see Bierman (2001, 2010).[37] Deutsche Bank (2014) reports modest long-term buyback premia. In a recently published study, Fu and Huang (2016) studied 14,309 stock repurchases over the 1984–2012 period and reported that long-run abnormal returns from stock repurchases for the 2003–2012 period became negative. We find no such results with corporate exports. Moreover, we find no such results over the 2003–2014 time period with global firms' stock buybacks.

Brealey, Myers, and Allen (2006) make an excellent analysis of the firm payout ratio, stock repurchases, and the financial decisions of corporations.

Debt is issued to finance capital expenditures and dividend payment.[38] Retained earnings add to the equity of the common shareholders and represent the majority of funds that can be used to finance additions to the operating assets of the corporation or to retire debt. An increase in assets financed by ownership capital as opposed to debt improves the credit standing of the firm and enables it to acquire debt funds at a relatively lower rate. Debt and internally generated funds are the primary sources of aggregate investment outlays.

New debt issues exceed new equity issues by a multiple exceeding eight times, a result consistent with Dhrymes and Kurz (1967); Guerard, Bean, and Andrews (1987); and Brealey, Myers, and Allen (2006). New debt issued is identified with the financing capital expenditures in the simultaneous equation modeling. Funds represented in earnings should increase the future profits of the shareholders and eventually result in buy backs or higher dividends. It is not the increment in the book value of the shares, but a hoped-for sequence of increased earnings that makes retained earnings of value to the shareholder. The doctrine of corporate exports is consistent with the financial planning structure to maximize the market value of the firm. The primary advantage of corporate exports is its use as a portfolio constraint to enhance portfolio returns, see Chapter 5.

Let's define a variable to designate the net corporate export of funds (CE) of the corporate sector, as defined in Guerard and Schwartz (2007) and Guerard et al. (2014).

$$CE = \text{Dividends Paid} + \text{Interest Paid} + \text{Net Equity Repurchased} - \text{Net Debt Issued} \quad (6)$$

In 1972, the corporate sector exported over $32 billion of funds and by 2014, the corporate sector funds exported to grown to over $ 145 billion. See Table 2.9. The surplus of funds over any possible reasonable capital investment policy is the rationale behind the cash buy back of shares and the payment of dividends. In short, the US corporate sector continues to be a net exporter of funds to the rest of the economy and has risen almost consistently throughout the 1972–2014 period. The steady growth of dividends paid and equity repurchased has led to a great growth in corporate exports.[39] IBM's corporate exports have risen substantially with its dividend payments, and a great reduction in common stock due to equity repurchases. Equity repurchases have substantially exceeded dividends for IBM in the post-2000 time period. Why should the reader care about corporate exports? Firms that engage in stock repurchases, pay dividends, and minimize net debt issuance, produce higher stocker returns than firms producing smaller levels of exports. We report these portfolio returns in Chapter 5.

Table 2.9: Corporate Exports Components, $MM

Year	Number of Companies	Long-Term Debt Issued	Long-Term Debt Repurchased	Stock Dividends	Equity Repurchased	Equity Issued	Net New Debt	Corporate Exports
			All Compustat Companies and IBM, Selected Years					
1972	4508	35435.7	20728.3	30444.9	2113.7	8679.8	14707.4	32123.1
1975	6757	93832.6	47912.0	41668.1	15238.0	18180.1	459320.6	28349.7
1980	6889	173022.6	82885.2	82523.2	8295.3	42367.2	90137.4	76272.3
1987	9225	437062.9	329071.9	154007.9	71264.0	102673.0	107991.1	250345.3
1990	9571	666371.5	512349.7	192196.8	55208.9	57453.2	154021.1	397666.3
1995	12491	1231890.1	898656.4	263986.6	96141.0	132137.5	333236.6	277587.0
2000	12092	2724008.9	1721730.0	367292.3	247253.7	411075.7	1002281.5	22446.5
2008	10884	6085805.7	5138659.0	776022.6	703572.4	891732.3	947147.0	954624.8
2010	11088	6036302.0	7142278.4	727427.8	521110.3	541200.9	-1105976.4	26009941.2
2014	10921	6703998.7	6000589.7	979835.3	824651.0	497677.5	703409.0	1453053.5
1972	IBM	201.6	98.3	626.2	0.0	270.6	103.3	37.4
1975	IBM	29.7	870.4	968.9	0.0	284.5	-40.7	787.4
1980	IBM	604.0	94.0	484.0	484.0	422.0	510.0	1885.0
1987	IBM	408.0	719.0	2654.0	1425.0	133.0	-311.0	4874.0
1990	IBM	4676.0	4184.0	2774.0	491.0	135.0	1592.0	2874.0
1995	IBM	6636.0	9460.0	634.0	10.0	0.0	-4110.0	4346.0
2000	IBM	9604.0	756.0	929.0	6093.0	0.0	2043.0	5696.0
2008	IBM	4363.0	3522.0	1250.0	6506.0	3774.0	841.0	7676.0
2010	IBM	8055.0	6522.0	3177.0	15376.0	3774.0	1533.0	14174.0
2014	IBM	8180.0	4644.0	4265	13679.0	709.0	3536.0	14729.0

Source: WRDS Database

Is IBM a unique company? We can use FactSet Investment Systems to download financial information for well over 46,500 firms.[40] The ratios are reported and produced on a consistent basis so that comparisons on profitability, leverage, efficiency, and composite analyses are valid for 20 years. We summarize 10 of the ratios used in Guerard and Schwartz (2007) in Table 2.10 for IBM, Dominion, and Facebook. In terms of basic profitability, as measured by the Return on Assets, IBM has the highest profitability. IBM's extensive leverage, as measured by the TDTA and reported in Table 2.10, leads to the highest ROE of the three firms. Facebook has the greatest liquidity, as measured by the Current and Quick Ratios, and the least leverage, as measured by TDTA. Is IBM a unique company? No. The growth of IBM sales has been essentially zero for the past 19 years, whereas Facebook Sales have grown at 78 percent annually for the past nine years, Only Dominion Energy, a utility firm with a 2 percent annualized growth rate is close to IBM's sales growth rate. IBM's low growth, extensive leverage, and high profitability make it a very interesting study.

Table 2.10: Comparative Factset Ratios, 5-Year Averages, 2011–2016

Ratio	IBM	Dominion	Facebook
ROE	86.11	12.21	10.30
ROA	12.52	2.70	9.08
ROIC	30.93	4.39	10.10
P/E	11.83	39.51	587.62
P/B	10.27	3.28	6.55
P/CF	10.02	10.32	32.41
TIE=EBIT / I	30.18	3.04	360.45
Long-Term Debt / Total Assets	27.17	39.42	3.00
TDTA	33.22	47.47	3.90
CR	1.22	0.67	11.08
Quick Ratio	1.17	0.49	11.08

2.10 Summary and Conclusions

The financial community has long calculated ratios to assess the liquidity, profitability, leverage, and efficiency of firms. Ratio calculations and analysis summarize financial information found on the balance sheet and income statement. The investor can often easily assess the financial health of a firm by calculating the ratios introduced in this chapter. Ratio analysis can be extremely useful in screening potentially poorer performing stocks, identifying problem firms, if not, bankrupt firms. The reader must be careful to compare a firm's ratios with its competitor and other firms in the economy. The reader is introduced to the concept of corporate exports, a variation on the sources and uses of funds that incorporates dividend payments, stock and debt issuance and buybacks. Corporate exports will be shown in Chapter 5 to enhance portfolio returns.

---

[1] Estimated from the *Statistical Abstract of the United States*, 2010, US Department of Commerce, Bureau of Census.

[2] There is a variant of the partnership known as the limited partnership where one or more partners may limit their liability to their investment in the firm. There still must be at least one general partner.

[3] Unless stipulated otherwise, losses are shared on the same basis as profits. If there is no express agreement, profits and losses are shared equally by the partners.

[4] Legally any payment of interest or salaries to the partners is not an expense but merely a way of distributing the partnership profits. However, these are generally subtracted on the accounting statements as an operating expense in order to obtain a pro forma profit figure.

[5] The heir of the deceased partner does not automatically become a member of the firm. If the surviving partners wish, they might let him in, actually forming a new partnership. Otherwise, the heir is entitled to a dissolution and accounting. It is quite usual, however, for the payment to the heir to be pre-set in the partnership agreement. The difficulties that can occur because of the death of a partner give a partnership an insurable interest in the life of its principals. The insurance benefits can help in settling the partnership accounts.

[6] The partnership is very common in smaller companies in the marketing field, in small manufacturing firms, in the professions, and in agriculture. The partnership form is very suitable where the personal skills and reliability of the partners is more important than capital in engendering income for the firm. Thus, doctors, lawyers, accountants, etc., have often been quite successful practicing as partners.

[7] Unless, as sometimes happens in a closely held corporation, the major shareholder personally endorses the corporation's note in order to obtain additional credit, or he has failed to pay in the full par value of the stock. This last possibility is not too likely in practice.

---

[8] Rights such as these can usually be voluntarily abridged. Thus, a family-held corporation can require that a stockholder offer his shares to the corporation or other existing stockholders (at some fixed price) before he can sell them to someone else.

[9] In exercising independent choice in the appointment of officers, most boards resemble the electoral college since they are already pledged to a given set of candidates. In the rare instances of a contest for control, the management runs one slate of directors and the dissidents nominate an opposing slate.

[10] Under the present tax laws, some closely held corporations that satisfy the requirements might elect to be taxed as partnerships. It depends on the individual situation whether this is advantageous.

[11] Estimated from the *Statistical Abstract of the United States*, 2010, US Department of Commerce, Bureau of Census.

[12] The reader probably remembers that 2008 was the year of the Global Finance Crisis (GFC). Corporate profits were $66 billion in 1970 for 1.655 million corporations; $239 billion in 1980 for 2.711 million corporations; $389 billion in 1990 for 3.72 million corporations; $928 billion in 2000 for 5 million corporations; and $1933 billion in 2000 for 5.8 million corporations.

[13] Analysts' expectations have been historically, since 1976 in the US and 1987 globally, more highly significantly associated with stockholder returns than reported earnings, which we will discuss in Chapters 5 and 6.

[14] *The Economic Report of the President*, February 2018, pp. 32-33.

[15] To refer to depreciation charges as a source of funds is the common shorthand usage, which to be frank is somewhat inaccurate. Strictly speaking, only the firm's operations provide funds. If the firm's revenues did not exceed its direct expenses, there would be no funds flow for the period. However, since the depreciation charge is added to retained earnings for the period to obtain the total of reinvested internally generated funds, the custom has grown of referring to depreciation as a source of funds.

[16] The reduction in the corporate tax rate in December 2017 increased net income by at least 14 for many companies, which will proportionally increase their market value. The United States Stock Market, as measured by the Standard & Poor's 50 Index, has appreciated some 37.78 percent from November 2016, when President Trump was elected, until the present time, January 26, 2018. Even (Southern, New-Deal) Democrats enjoy stock market profits.

[17] Essentially unrealized profits are ignored but unrealized losses affect the accounting results.

[18] White, Sondhi, and Fried (1998), pp. 274–275.

[19] White, Sondhi, and Fried (1998), pp. 289–290.

[20] This constitutes "historical cost depreciation" in contrast to "replacement cost depreciation," where allowance might be made for changes in current costs.

[21] On the other hand, a fall in price levels would enable a firm to increase its physical capacity through the reinvestment of depreciation allowances.

[22] Sloan (1996) and Richardson, Sloan, Soliman, and Tuna (2001) have shown that accruals provide information about earnings' quality. Inventory accruals, the primary source of asset accruals, and accounts payables, the primary source of liability accruals, convey information. Richardson, Sloan, Soliman, and Tuna (2001) showed that the setting of accruals, asset accruals less liability accruals, is an important source of financial information. Nondiscretionary accruals, accruals associated with sales growth and the level of firm operating activities, conveys information about earnings quality. Accrual information influences information attributable to efficiency. Sloan uses FASB 95 to define accruals as the difference between net income and cash flows from operating activities. Sloan (1996) found that stocks with extreme accruals have less persistent earnings and experience mean reversion in next year's earnings.

[23] They might be called the proprietorship account in a single proprietorship or the partner's equity in a partnership.

[24] In bank balance sheets, the equivalent to the unappropriated earned surplus account is titled undivided profits. The bank's surplus account consists of capital surplus plus the retained earnings that are considered permanently committed to the business.

[25] To add to the beginner's confusion, current operating losses are often called operating deficits, or just deficits.

[26] Modigliani and Miller, Proposition II, see Lerner and Carleton, *A Theory of Financial Analysis* (1966).

[27] Operating statements for internal control can be made up for any feasible time period. Most large firms present quarterly statements for their investors, although the annual results carry the most weight and go into the record books of the financial services, such as *Mergent*'s (formerly *Moody*'s), *Standard & Poor's*, et. al.

---

[28] That is, if the recipient of the income is an individual and not a nontaxable institution or another corporation. An 85 percent dividend received credit is given to corporations for dividends received from another corporation that is non-consolidated. (To consolidate for tax purposes, the parent must own a minimum of 80 per cent of the subsidiary's shares.) The other 15 percent is taxed at the usual corporate income tax rate.

[29] Drtina and Largay (1985) examined cash flow reporting with regard to "cash flow from operations" (CFO). Problems can develop because of ambiguity in terms of the definition of "operations," the measurement of the current position of long-term leases, diversity in reporting practices, and reclassification of current and noncurrent accounts. Krishnan and Largay (2000) reported that past direct method cash flows offer better predictions of future operating cash flows than indirect method cash flows information. The direct method of presenting cash flows OCF for time t is predicted to be:

$OCF_t = \int(CSHRD_{t-1}, CSHPD_{t-1}, INTRD_{t-1}, INTPD_{t-1}, TXPD_{t-1})$

where

$CSHRD_{t-1}$ = Cash Received from Customers at time t-1,

$CSHPD_{t-1}$ = Cash Paid to Suppliers and Employees at time t-1,

$INTRD_{t-1}$ = Interest Received at time t-1,

$INTPD_{t-1}$ = Interest Paid at time t-1,

and

$TXPD_{t-1}$ = Taxes Paid at time t-1.

[30] Guerard and Schwartz (2007) used IBM, DuPont, and Dominion Resources, a large, Southern utility firm as example firms for examining income statement, balance sheet, and sources and uses of cash flow. One of the authors, Guerard, has owned Dominion Resources since 1982 and enjoyed a very substantial increase in personal wealth, outperforming the S&P 500 by over 150 basis points with much less risk (a lower beta).

[31] This means that these companies, in effect, carry no net working capital.

[32] The turnover figure, however, cannot necessarily be improved before the event by raising markups, since such an action might reduce total sales. The insights of economic theory are useful at this point. The effect of the markup on the sales/inventory ratio depends on the relative inelasticity of the firm's demand.

[33] The larger the percentage of ownership capital in the financial structure, the less will be any difference between rate of net profit on equity and the rate of net profit on total assets.

[34] Intangible assets, such as good will, should be subtracted from the book value of the common equity.

[35] Pierre DuPont and Donald Brown, a DuPont employee, developed the DuPont return on investment relationship to access the firm's financial performance. General Electric calculated profitability by dividing earnings by sales (or costs). However, this calculation ignored the magnitude of invested capital. In 1903, Pierre DuPont created a new general ledger account for "permanent investment," where capital expenditures were charged at cost. The DuPont Corporation executive committee was presented with monthly sales, income, and return on invested capital on the firm's thirteen products in 1904 (Chandler (1977)). Donald Brown contributed to the DuPont analysis pointing out that as sales volume rose, the return of invested capital rose, even if prices remained constant. Brown's "turnover" analysis was defined as sales divided by total investment.

[36] Lakonishok and Vermalen reported premiums of 21.79%, 24.09%, and 18.54% on tender offers during the 1962–1986, 1962–1979, and 1980–1986 periods, respectively. They also reported cumulative abnormal returns of 12.54%, 14.58%, and 9.78% to non-tendering stockholders during the corresponding periods. Smaller firms produced the highest abnormal returns.

[37] A forecast of dividend levels can be obtained by studying past payout rates and attempting to predict future earnings levels. The amount left after common dividends are paid represents retained earnings or earnings reinvested in the firm. Firms target a target debt ratio by choosing between debt and equity repurchases rather than choosing between debt and equity issuances, see Bierman (2001). Repurchases and dividends are motivated by contracting future investment opportunities, see Bierman (2010).

[38] Dhrymes and Kurz (1967) proposed an explicit link among these dividend, investment, and debt issuance decisions and econometrically implemented using sample consists of 181 industrial and commercial firms for which a continuous record exists over the period 1947–1960. Guerard, Bean, and Andrews (1987) updated the Dhrymes and Kurz analysis and reported: (1) a strong interdependence is evident between the investment and dividend decisions and (2) a strong interdependence is evident between the investment and new debt financing decisions.

[39] An analysis of the CE components reveals that interest paid has risen faster than dividends paid during the 1971–2014 time period. Net debt issues have risen at an undiminished rate, with the notable exception of 2001–2005 and 2009. Stock repurchases rose substantially following the crash of October

1987. Net equity repurchases increased substantially in the 2002–2006 time period and fell dramatically with the financial crisis. To illustrate the corporate fund generation process, Guerard (2010) showed the ten largest and smallest corporate exporter firms in 1983; the largest corporate exports firms that included AT&T, IBM, and several of the large oil companies dominated positive corporate exports in 1983 as they paid large dividends and interest and generally re-purchased more debt than was issued (which made a great deal of sense given the level of interest rates in 1983). A similar process occurred in 2006 as Microsoft, Pfizer, and the oil companies dominated the largest corporate exporting firm (IBM fell to only the 24th largest exporter in 2006). Dividends paid exceeded equity repurchased of the Compustat firms from 1971–2014, although equity repurchases have risen relatively to dividends since 1982. We examine the 1971–2014 period because Compustat does not maintain debt and equity issuance, and repurchases, prior to 1971. Guerard et al. (2014) created mean-variance efficient portfolios and tilted the portfolios to purchase stocks with the largest corporate exports. We believe, and will show, that buying stocks on a forecasted earnings acceleration variable basis can be complemented with buying companies that generate positive and corporate exports. The portfolios produced statistically significant out-performance relative to the universe benchmark. We report that portfolios with the largest corporate export stocks produced higher returns, relative to risk, when the stocks that were not identified as under-valued by a stock selection model were excluded from the portfolios.

[40] We download income statement, balance sheet, and cash flow forecasts for that number of stocks every Tuesday evening at McKinley Capital Management, LLC, to produce our weekly database.

Chapter 3: The Risk and Return of Equity and the Capital Asset Pricing Model

3.1 Introduction

Individual investors must be compensated for bearing risk. It seems intuitive that there should be a direct linkage between the risk of a security and its rate of return. We are interested in securing the maximum return for a given level of risk, or the minimum risk for a given level of return. The concept of such risk-return analysis is the efficient frontier of Harry Markowitz (1952, 1959). If an investor can invest in a government security that is backed by the taxing power of the Federal Government, then that government security is relatively risk-free. The 90-day Treasury bill rate is used as the basic risk-free rate. Supposedly the taxing power of the Federal government eliminates default risk of government debt issues. A liquidity premium is paid for longer-term maturities, due to the increasing level of interest rate risk. Investors are paid interest payments, as determined by the bond's coupon rate, and can earn market price appreciation on longer bonds if market rates fall or endure losses if market rates rise. During the period from 1928–2017, as mentioned in Chapter 1, Treasury bills returned 3.44 percent, longer-term government bonds earned 5.15 percent, and corporate stocks, as measured by the stock of the S&P 500 index, earned 11.53 percent annually. The annualized standard deviations are 3.44, 5.15, and 19.62 percent, respectively, for Treasury bills, Government Bonds, and stocks (S&P). The risk-return trade-off has been relevant for the 1928–2017 period. Why do corporate stocks offer investors such returns?

First, stockholders own a fraction, a very small fraction for many investors, of the firm. When one owns stocks, one is paid a dividend and earns stock price appreciation. That is, an investor buys stock when he or she expects its stock price to rise and compensate the investor for bearing the risk of the stock's price movements. Investors have become aware in recent years that not all price movements are in positive directions. It is common sense to view the negative returns as risk. Linking negative returns to a coherent and quantifiable risk measure is the foundation for the development of modern portfolio theory started by Markowitz.

The discussion thus far has been concerned with owning stock, or being "long" on the stock. Not all investors want to own stock. Imagine an investor who hates her mobile phone. The investor does not want to hear the "beep" of a text or email. The investor prefers to watch basketball without the distraction of a ringing mobile phone. The investor wants all communications firms to go bankrupt. In such a case, the investor could short a stock, borrowing from a broker, and purchasing the stock in the future. Suppose the investor saw that the price of AT&T, Inc, was $30 per share on February 1, 2019. The investor calls her broker and says short 200 shares of AT&T. If the price of AT&T falls to $12.25, the investor profits $3550; that is, ($30 - $12.25) *200 shares, or 59.2 percent. The investor can have peace and quiet, stock trading profits, and can watch basketball, "Top Chef", "The Blacklist" or other television shows without telephone interruptions.

Suppose an investor truly believed that the current economic and stock market boom will cease and the stock market will collapse. In such a case, the investor could sell all his stocks at their relatively high levels, and buy puts (options to sell stock), or sell stocks short, profiting handsomely if the Dow Jones Industrial Average falls from 25,000 to 15,000–18,000.

3.2 Calculating Holding Period Returns

A very simple concept is the holding period return (HPR) calculation, in which one assumes that the stock was purchased at last (period's) year's price and the investor earns a dividend per share for the current year and a price, appreciation (depreciation) relative to last year's price.

$$HPR_t = \frac{D_t + P_t - P_{t-1}}{P_{t-1}}$$

where D_t = current year's dividend

P_t = current year's stock price

P_{t-1} = last year's stock price

HPR_t = current year's holding period return

The assumption of annual returns is arbitrary, but common in finance. Markowitz illustrates portfolio theory in his seminal text, *Portfolio Selection*, using the annual returns data in Chapter 2. It makes it easier to compare the returns of stocks with the returns of cash and bonds whose interest rates and coupon rates are quoted in annual basis. Most literature, for example, Weston and Brigham (1980), uses annual return to illustrate risk-return tradeoffs in the modern portfolio theory like Markowitz did in 1959. We list the recent closing price and dividends in Table 3.1 for Dominion Resources (ticker symbol: D) and International Business Machine (ticker symbol: IBM).

Table 3.1: Annual Dividends and Prices, 2007–2017

	IBM		D	
Year	**Ending Price**	**Dividends**	**Ending Price**	**Dividends**
2007	108.1	1.5	47.45	2.525
2008	84.16	1.9	35.84	1.58
2009	130.9	2.15	38.92	1.75
2010	146.76	2.5	42.72	1.83
2011	183.88	2.9	53.08	1.97
2012	191.55	3.3	51.8	2.11
2013	187.55	3.7	64.69	2.25
2014	160.44	4.25	76.9	2.4
2015	137.62	6.3	67.64	2.59
2016	165.99	6.9	76.59	2.8
2017	153.12	5.9	81.06	3.035

We calculate the IBM and D holding period returns in Table 3.2.

Table 3.2: Holding Period Returns, 2008–2017

Year	IBM			D		
	Total	**Price**	**Dividends**	**Total**	**Price**	**Dividends**
2008	20.39%	22.15%	1.76%	21.14%	24.47%	3.33%
2009	58.09%	55.54%	2.55%	13.48%	8.59%	4.88%
2010	14.03%	12.12%	1.91%	14.47%	9.76%	4.70%
2011	27.27%	25.29%	1.98%	28.86%	24.25%	4.61%
2012	5.97%	4.17%	1.79%	1.56%	-2.41%	3.98%
2013	-0.16%	-2.09%	1.93%	29.23%	24.88%	4.34%

Year	IBM			D		
2014	12.19%	14.45%	2.27%	22.58%	18.87%	3.71%
2015	10.30%	14.22%	3.93%	-8.67%	12.04%	3.37%
2016	25.63%	20.61%	5.01%	17.37%	13.23%	4.14%
2017	-4.20%	-7.75%	3.55%	9.80%	5.84%	3.96%

Unpredictable changes in annual ending prices are the source of variability in annual HPRs. For example, in 2009, the price of IBM rose from $84.16 in 2008 to $130.9. The $46.76 price appreciation and $2.15 dividends lead to an annual return of 58.09 percent.

$$IBM\ \ HPR_{2009} = \frac{\$2.15 + \$130.9 - \$84.16}{\$84.16} = .5809$$

The price movement of IBM in 2009 was much higher than Dominion Resources' 13.48 percent on a $35.84 base. Dominion Resources' holding period returns (HPRs) were consistently positive during the 2008–2017 period. The HPRs of stocks are often far more dependent upon stock price movements than the dividends received by investors. The HPRs of IBM range from 58.08 percent in 2009 to -20.39 percent in 2008. One can estimate an expected return for IBM by calculating the mean value of the annual HPRs during the 2008–2017 period. The expected return for IBM during the 2008–2017 period was 8.38 percent, with a standard deviation of 23.55 percent. If the annual HPRs of IBM are normally distributed, that is, the returns fall within the normal "Bell" curve, then 67.6 percent of the annual observations of IBM returns should fall within the -15.18 percent and 31.93 percent range (one standard deviation). The reader immediately sees how wide the one standard deviation range is for annual returns. As shown in Table 3.3, we can calculate the expected returns for Dominion Resources in a similar manner. The returns are extremely volatile for IBM, having a standard deviation of 23.55 percent. It might be worthwhile to calculate a coefficient of variation (CV) in which the standard deviation is divided by the expected return. The calculation of the CV leads the investor to recognize that IBM, more than Dominion Resources, produces greater variation for a given level of expected returns. Note that the calculations of expected returns and standard deviations allow the investor to allocate scarce resources on the basis of historic returns and risk.

Table 3.3: Expected Return and Standard Deviations

Security	Ticker	E(R)	σ	CV
Dominion Resources	D	10.75%	16.17%	1.50
IBM	IBM	8.38%	23.55%	2.81

An investor should not allocate resources to only one security, as was the case with many Enron stockholders. Remember the expression, "Do not put all of your eggs in one basket." Clearly the standard deviation of return might be reduced by investing in several assets, particularly if these assets are somewhat uncorrelated. An investor does not benefit from investing in two stocks, as opposed to only one security, if both stocks move in parallel. That is, if stock A rises 10 percent and stock B rises 10 percent, then it is not evident that the investor has any benefits to a second stock investment. However, if Dominion Resources has an expected return of 10.75 percent and IBM has an expected return of 8.38 percent, an investor can purchase an equal dollar amount of each stock and reduce risk, if the stocks are not perfectly correlated with each other. The correlation coefficient is the covariance of two series divided by the product of the respective standard deviations. The correlation coefficient allows an investor to understand the statistical nature of a covariance because the correlation coefficient is bonded between -1 and +1. Low correlation coefficients imply that the assets are not good substitutes for one another, and diversification is enhanced by using assets with lower correlations.

Let us turn to a more modern source of financial data: the stock tapes of the CRSP database, which are monthly returns from the Center for Research in Security Pricing (CRSP) data. The CRSP database, created by Lawrence Fisher and James Lorie, at the University of Chicago, starting in 1960, collected monthly prices and dividends data on all NYSE and AMEX stocks, 1926 to 1962. Data analysis became infinitely easier, leading initially to classic studies by Fama (1965), Fisher (1966), and Fama, Fisher, Jenson, and Roll (1969), which established most of the basis of market efficiency (1970).

3.3 Markowitz on Portfolio Risk

Assume you have two securities to invest whose future returns are denoted by random variables R_1 and R_2. A portfolio is a linear combination of these two assets, and the portfolio's future return, a random variable R_p, is linked to the security one and two's return variable by

$$R_p = x_1 R_1 + x_2 R_2 \tag{1}$$

Where x_1 and x_2 are the weights assigned to security one and two. They must satisfy the budget constraint

$$x_1 + x_2 = 1 \tag{2}$$

In finance literature, the expected value of a return variable R is called expected return. Markowitz (1952) defines the standard deviation of a return variable R as risk. For computation convenience, it is always best to compute the portfolio's variance first, then take the square root of it to get standard deviation or risk.

In the two-security case, the portfolio return and risk are given by

$$\mathrm{E}(\mathrm{R}_p) = x_1 \mathrm{E}(\mathrm{R}_1) + x_2 \mathrm{E}(\mathrm{R}_2) \tag{3}$$

and

$$\sigma_p^2 = \mathrm{var}(R_p) = \mathrm{var}(x_1 R_1 + x_2 R_2) = x_1^2 \sigma_1^2 + x_2^2 \sigma_2^2 + 2 x_1 x_2 \sigma_{12} \tag{4}$$

where

σ_1, σ_2 are the standard deviation of assets 1, 2

σ_{12} = covariance of assets 1, 2

$\sigma_{12} = \rho_{12} \sigma_1 \sigma_2$

ρ_{12} = correlation coefficients of assets 1, 2

We have calculated the means and standard deviation for securities IBM and D. In order to calculate the portfolio risk, equation (4) shows that we also need to calculate the covariance or correlation of returns. We can use the CORR procedure in SAS to do all the calculations.

Program 3.1: Correlation of IBM and Dominion Resources

```
DATA IBM_D_return_data;
INPUT IBM D;
cards;
   20.39         21.14
   58.09         13.48
   14.03         14.47
   27.27         28.86
   5.97          1.56
   -0.16         29.23
   12.19         22.58
   10.30         -8.67
   25.63         17.37
   -4.20         9.80

;
proc corr data = IBM_D_return_data;
 var IBM D;
run;
```

Output 3.1: Return Correlation of IBM and Dominion Resources

The SAS System

The CORR Procedure

2 Variables: IBM D

Simple Statistics

Variable	N	Mean	Std Dev	Sum	Minimum	Maximum
IBM	10	8.37500	23.55650	83.75000	-20.39000	58.09000
D	10	10.75400	16.17273	107.54000	-21.14000	29.23000

Pearson Correlation Coefficients, N = 10
Prob > |r| under H0: Rho=0

	IBM	D
IBM	1.00000	0.44512 0.1974
D	0.44512 0.1974	1.00000

It shows the correlation of annual returns reported by Table 3.2 is

$$\rho_{12} = \rho_{IBM,D} = 0.4451$$

In an equally weighted portfolio, $x_1 = x_2 = .50$. Let x_1 = wt. of IBM and x_2 = wt. of D. The portfolio expected return is a weighted combination of asset expected returns as equation (3).

$$E(R_{EQ}) = .5(8.38\%) + .5(10.75\%) = 4.19\% + 5.38\% = 9.56\%.$$

The portfolio's variance is weighted sum of variance and covariance by equation (4).

$$\sigma^2(R_{EQ}) = (.5)^2(23.55\%)^2 + (.5)^2(16.17\%)^2 + 2(.5)(.5)(.4451)(23.55\%)(16.17\%)$$

$$= 0.0139 + .0065 + .0085 = .0289$$

Therefore, the risk of the equally weighted portfolio is

$$\sigma(R_{EQ}) = \sqrt{0.0289} = 16.99\%$$

The annual expected return on an equally weighted portfolio of Dominion Resources and IBM stock is 9.56 percent, and its corresponding standard deviation is 16.99 percent. The portfolio return should fall within the range of -24.42 percent and 43.54 percent approximately 95 percent of the time. This range corresponds to the expected return plus or minus two standard deviations of return. The corresponding one standard deviation of returns range is -7.43 percent and 26.53 percent, respectively. By varying the portfolio weight subject to budget constraints (2), we can trace out all the possible risk and return combinations as shown in Figure 3.1.

Figure 3.1 Two-Securities Risk-Return Curve

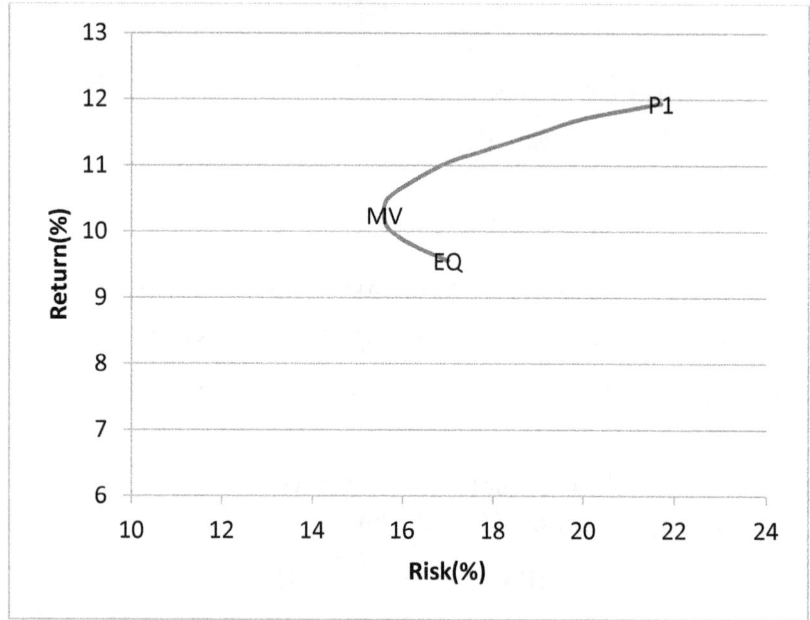

The risk return curve is parabolic. The portfolio corresponding to the lower part of the parabola MV to EQ is not desirable because there are portfolios corresponding to the upper part of the parabola MV to EQ such that they have more returns and the same risk.

The diversification reduces risk but cannot get rid of risk as shown in Figure 3.1. The portfolio corresponding to the label MV is the portfolio that has minimum risk between these two stock combinations. One can find this portfolio analytically. To minimize risk, one seeks to allocate resources to assets such that the change in risk goes to zero with the change in the percentage invested in the asset. That is, risk minimization implies a first partial derivative of being zero, with respect to the change in the asset weight. From budget equation (2), $x_2 = 1 - x_1$, the variance of the portfolio by equation (4) becomes

$$\sigma_p^2 = x_1^2\sigma_1^2 + (1-x_1)^2\sigma_2^2 + 2x_1(1-x_1)\sigma_{12}$$

$$= x_1^2\sigma_1^2 + (1-x_1)^2\sigma_2^2 + 2x_1(1-x_1)\rho_{12}\sigma_1\sigma_2$$

$$= x_1^2\sigma_1^2 + (1-x_1)(1-x_1)\sigma_2^2 + 2x_1\rho_{12}\sigma_1\sigma_2 - 2x_1^2\rho_{12}\sigma_1\sigma_2$$

$$= x_1^2\sigma_1^2 + (1-2x_1+x_1^2)\sigma_2^2 + 2x_1\rho_{12}\sigma_1\sigma_2 - 2x_1^2\rho_{12}\sigma_1\sigma_2$$

The risk minimizing portfolio should satisfy

$$\frac{\partial\sigma_p}{\partial x_1} = 0$$

Therefore,

$$\frac{\partial\sigma_p^2}{\partial x_1} = 2\sigma_p\frac{\partial\sigma_p}{\partial x_1} = 0$$

Which implies

$$\frac{\partial \sigma_p^2}{\partial x_1} = 2x_1\sigma_1^2 - 2\sigma_2^2 + 2x_1\sigma_2^2 + 2\rho_{12}\sigma_1\sigma_2 - 4x_1\rho_{12}\sigma_1\sigma_2 = 0 \tag{5}$$

Moving the constants to the right-hand side leads to

$$2x_1\sigma_1^2 + 2x_1\sigma_2^2 - 4x_1\rho_{12}\sigma_1\sigma_2 = 2\sigma_2^2 - 2\rho_{12}\sigma_1\sigma_2$$

$$(\sigma_1^2 + \sigma_2^2 - 2\rho_{12}\sigma_1\sigma_2)x_1 = \sigma_2^2 - \rho_{12}\sigma_1\sigma_2$$

$$x_1 = \frac{\sigma_2(\sigma_2 - \rho_{12}\sigma_1)}{\sigma_1^2 + \sigma_2^2 - 2\rho_{12}\sigma_1\sigma_2} \tag{6}$$

Equation (6) shows the risk-minimizing weight (percentage invested) of asset one in the portfolio.

If an investor seeks to minimize risk with a two-asset portfolio with IBM as asset one and D as asset two, then the optimal investment in asset one, IBM is given by equation (6).

$$x_1 = \frac{.1617(0.1617 - 0.4451 * 0.2355)}{0.2355 * 0.2355 + 0.1617 * 0.1617 - 2 * 0.4451 * 0.1617 * 0.2355}$$

$$= \frac{.0092}{.0477}$$

$$= .19$$

By budget equation (2),

$x_2 = 1 - .19 = 0.80$

An investor should invest 20 percent of the portfolio in IBM and 80 percent in Dominion Resources.

3.4 An Introduction to Modern Portfolio Theory

Markowitz created a portfolio construction theory in which investors should be compensated with higher returns for bearing higher risk. The Markowitz framework measured risk as the portfolio standard deviation, its measure of dispersion, or total risk. The Sharpe (1964), Lintner (1965), and Mossin (1966) development of the Capital Asset Pricing Model (CAPM) held that investors are compensated for bearing not total risk, but rather market risk, or systematic risk, as measured by the stock beta. An investor is not compensated for bearing risk that can be diversified away from the portfolio. The beta is the slope of the market model, in which the stock return is regressed as a function of the market return.

The simplest non-trivial portfolio choice problem discussed in previous section shows that some portfolios are "good," and some portfolios are "bad" judged on risk-return basis. Markowitz defines the "good" portfolios as the efficient portfolios. For an efficient portfolio, there is no other portfolio that could have more returns with the same risk or less risk with the same return. The risk-return curve generated by efficient portfolios, whether it is formed by two securities or thousand securities, is called the efficient frontier and it is an upper forwarding parabola-like curve. By definition, there should no portfolio on the upper left side of the efficient frontier. As the risk of the portfolio rises, its expected return must rise.

In case of n investment securities, a portfolio's expected return is

$$\mu_p = \sum_{i=1}^{n} w_i \mu_i \tag{7}$$

And the portfolio's variance is

$$V_p = \sum_{i=1}^{n} \sum_{j=1}^{n} w_i w_j c_{ij} \tag{8}$$

where $w' = (w_1, w_2, ..., w_n)$ is the portfolio weights satisfying budget constraint;

$$\sum_{i=1}^{n} w_i = 1 \tag{9}$$

and $\mu' = (\mu_1, \mu_2, ..., \mu_n)$ is the expected return vector, and $C = \{c_{ij}\}$ is the variance and covariance matrix. The efficient frontier can be found by quadratic programming. The general form of the Markowitz portfolio construct method is

$$\min_w V_p \tag{10}$$

Such that

$$\mu_p = E_p \tag{10a}$$

$$Aw = b \tag{10b}$$

$$w \geq 0 \tag{10c}$$

Where A is an m×n matrix, and b is an m component vector. Constraints (10c) assume an investor cannot short-sell securities (i.e., borrow security from another investor to sell). We recommend Markowitz's (1959) critical line algorithm to compute efficient frontier because of its efficiency and because it does not require the variance-covariance matrix to be invertible.

In case there are no short-selling constraints (10c), the only constraint (10b) is the budget constraint (9), and the variance-covariance matrix is invertible, the efficient frontier can be expressed analytically as derived in equation (12) of Chapter 1.

$$\sigma_p^2 = (\frac{E_p \ell' C^{-1} \ell - \mu' C^{-1} \ell}{\mu' C^{-1} \mu \ell' C^{-1} \ell - (\mu' C^{-1} \ell)^2})^2 \mu' C^{-1} \mu + (\frac{\mu' C^{-1} \mu - E_p \mu' C^{-1} \ell}{\mu' C^{-1} \mu \ell' C^{-1} \ell - (\mu' C^{-1} \ell)^2})^2 \ell C^{-1} \ell$$

$$+2 \frac{E_p \ell' C^{-1} \ell - \mu' C^{-1} \ell}{\mu' C^{-1} \mu \ell' C^{-1} \ell - (\mu' C^{-1} \ell)^2} \frac{\mu' C^{-1} \mu - E_p \mu' C^{-1} \ell}{\mu' C^{-1} \mu \ell' C^{-1} \ell - (\mu' C^{-1} \ell)^2} \mu' C^{-1} \ell \tag{11}$$

The efficient frontier derived from equation (11) is shown in Figure 3.2 labeled with diamonds.

Figure 3.2: Efficient Frontier

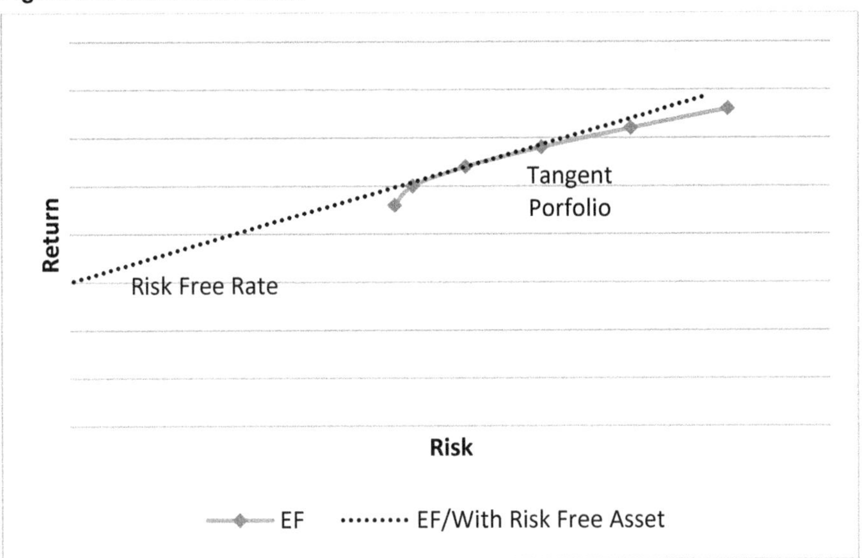

When there is a risk-free asset that an investor can borrow and lend at the risk-free rate, R_F, then the risk-return efficient frontier becomes a line as pictured in Figure 3.2. This line passes through the risk-free rate and intersects with efficient frontier created without the risk-free asset at only one point, the tangent point.

Any risk–return on this line can be expressed by the slope of the tangent portfolio and intercept point

$$E(R_P) = R_F + \left[\frac{E(R_{TP}) - R_F}{\sigma_{TP}} \right] \sigma_P \tag{12}$$

$E(R_P)$ = Expected return on the portfolio;

and σ_P = standard deviation of the portfolio.

From investment point of view, equation (12) states that all investors will hold risk-free asset and tangent portfolios only. Investors will not invest in any other efficient portfolios composed of securities without risk-free assets.

By definition, the slope of the tangent portfolio's risk-return line is maximized as

$$\frac{E(R_{TP}) - R_F}{\sigma_{TP}} = \max_{\{w'\}} \frac{E(R_p) - R_F}{\sigma_p} \tag{13}$$

with budget constraint (9). Let $(w_{TP})' = (w_1^{TP}, w_2^{TP}, \ldots, w_n^{TP})$ the portfolio weight of the tangent portfolio, then it will satisfy the first order condition with Lagrange multiplier λ:

$$\frac{\partial}{\partial w_i}\left(\frac{E(R_p) - R_F}{\sigma_p}\right)\Bigg|_{w^{Tp}} - \lambda = 0, \text{ for i =1,2,..,n.} \tag{14}$$

Since

$$\frac{\partial}{\partial w_i}\left(\frac{\mathrm{E}(\mathrm{R}_p)-\mathrm{R}_\mathrm{F}}{\sigma_p}\right)=\frac{\mu_i}{\sigma_p}-\frac{\mathrm{E}(\mathrm{R}_p)-\mathrm{R}_\mathrm{F}}{(\sigma_p)^2}\frac{\partial\sigma_p}{\partial w_i}=\frac{\mu_i}{\sigma_p}-0.5\frac{\mathrm{E}(\mathrm{R}_p)-\mathrm{R}_\mathrm{F}}{(\sigma_p)^3}\frac{\partial\sigma_p^2}{\partial w_i}$$

$$=\frac{\mu_i}{\sigma_p}-\frac{\mathrm{E}(\mathrm{R}_p)-\mathrm{R}_\mathrm{F}}{(\sigma_p)^3}\left(\sum_{j=1}^{n}c_{i,j}w_j\right)$$

The first order condition (14) becomes

$$\frac{\mu_i}{\sigma_{TP}}-\frac{\mathrm{E}(\mathrm{R}_{TP})-\mathrm{R}_\mathrm{F}}{(\sigma_{Tp})^3}\left(\sum_{j=1}^{n}c_{i,j}w_j^{TP}\right)-\lambda=0,\ \text{for i=1,2,…n} \qquad (15)$$

Multiply the tangent portfolio weight w_i^{TP} to the ith equation of (15), then sum them up

$$\frac{\sum_{i=1}^{n}\mu_i w_i^{TP}}{\sigma_{TP}}-\frac{\mathrm{E}(\mathrm{R}_{TP})-\mathrm{R}_\mathrm{F}}{(\sigma_{TP})^3}\left(\sum_{i=1}^{n}w_i^{TP}\sum_{j=1}^{n}c_{i,j}w_j^{TP}\right)-\lambda\sum_{i=1}^{n}w_i^{TP}=0$$

With the definitions of portfolio expected return, variance, and budget equations (7), (8) and (9), we get

$$\frac{\mathrm{E}(\mathrm{R}_{TP})}{\sigma_{TP}}-\frac{\mathrm{E}(\mathrm{R}_{TP})-\mathrm{R}_\mathrm{F}}{(\sigma_{TP})^3}(\sigma_{TP})^2-\lambda=0 \qquad (16)$$

which gives us the Lagrange multiplier λ as

$$\lambda=\frac{\mathrm{R}_\mathrm{F}}{\sigma_{TP}} \qquad (17)$$

Substituting the Lagrange multiplier λ of (17) into first-order condition (15), and rearranging the terms, it becomes

$$\mu_i-\mathrm{R}_\mathrm{F}=(\mathrm{E}(\mathrm{R}_{TP})-\mathrm{R}_\mathrm{F})\frac{\sum_{j=1}^{n}c_{i,j}w_j^{TP}}{(\sigma_{TP})^2}$$

$$=(\mathrm{E}(\mathrm{R}_{TP})-\mathrm{R}_\mathrm{F})\frac{\mathrm{cov}(R_i,R_{TP})}{(\sigma_{TP})^2} \qquad (18a)$$

$$=(\mathrm{E}(\mathrm{R}_{TP})-\mathrm{R}_\mathrm{F})\frac{\mathrm{cov}(R_i-R_F,R_{TP}-R_F)}{(\sigma_{TP})^2} \qquad (18b)$$

Equation (18b) follows from (18a) because the risk-free rate has no risk. This is the security market line, which states that the expected returns of securities are linear related.

Since all expected returns, variance-covariance, even risk-free rate, are not observable, and cannot be constants for any period of time. The tangent portfolio cannot be calculated by optimizations. Endless debates conclude that the market portfolio is a good proxy of the tangent portfolio. With that assumption, the straight line of the efficient portfolio by equation (12) becomes

$$E(R_P) = R_F + \left[\frac{E(R_M) - R_F}{\sigma_M}\right]\sigma_P \tag{19}$$

Which is called Capital Market Line (CML). The CML line specifies how investors should invest when there is a risk-free rate asset.

With the assumption that the market portfolio is the tangent portfolio, the security market line by equation (18) becomes

$$\mu_i - R_F = (E(R_M) - R_F)\frac{\text{cov}(R_i - R_F, R_M - R_F)}{(\sigma_M)^2} = (E(R_M) - R_F)\beta_i \tag{20}$$

This is the Capital Asset Pricing Model, which states that the securities expected returns are linearly related to systematic risk, as measured by the beta.

Figure 3.3: Security Market Line

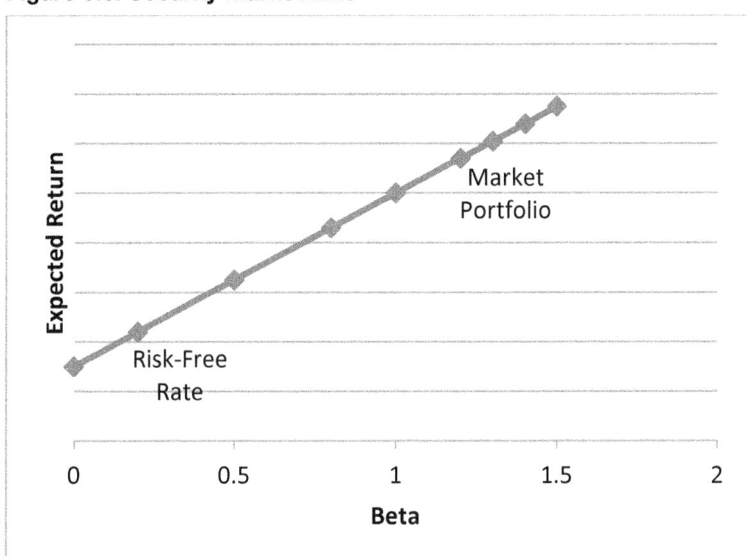

Since expected returns are not observable, CAPM of equation (20) suggests that any security's observed return should be linearly related to the observed return of market

$$R_{jt} = R_{Ft} + \beta_j[R_{Mt} - R_{Ft}] + e_{jt} \tag{21}$$

where

R_{jt} = security return at time t,

R_{Mt}= return on the market at time t,

R_{Ft} = risk-free rate at time t,

β_j= security beta, and

e_j = randomly distributed error term.

The difference of the security's return to risk-free rate is called the excess return. In this form, the β_j is the regression slope of the security's excess return against market excess return.

For any short estimation period, the variance of the observed risk-free rate is very close to zero, and we can treat the risk-free rate as constant. We can drop the subscript t for the risk-free rate in equation (21) to model observed security return as

$$R_{jt} = \alpha_j + \beta_j R_{Mt} + e_{jt} \tag{22}$$

In this form, the β_j is the regression slope of the security's return against market return. The regression model beta is the sole risk parameter of the Capital Asset Pricing Model.

3.5 Estimating Stock Betas

Let us estimate beta coefficients to be used in the Capital Asset Pricing Model (CAPM) to determine the rate of return on equity. One can regress monthly HPRs of Boeing, DuPont, and IBM against a value-weighted CRSP index, vwretd, which is an index of all publicly traded securities, or the S&P 500 Index, sprtrn. Most security betas are estimated using five years of monthly data, some sixty observations, although one can use almost any number of observations. We will estimate IBM, DD, and BA stock betas using OLS and Robust Regression for monthly observations, 2012-2016. We use the SAS robust regression procedure (PROC ROBUSTREG) Optimal Influence Function, OIF, with the Tukey weighting procedure for 99 percent to estimate the robust betas.[1]

We use Program 3.2 to estimate beta. There are three parts in this code. The first part reads CRSP monthly return data cross sectionally. The second part creates the time series returns for IBM and indices. The third part runs the OLS regression. The data creating part could be much simpler if the investor is interested in the estimation beta of one security and has returns in CVS file format.

Program 3.2: IBM Beta Estimation Against S&P500 Index

```
Title IBM Beta2016;
data CRSP1201;
infile "C:\QCF 2Ed\Data092018\SAS\Data\CRSP\CRSP1201.txt" lrecl=102;
input permno 1-6 cusip $ 8-15 @17 mcap f9.2 @27 ret f7.4 @35 vwretd f7.4 @43
sprtrn f7.4 @85 VOL f8.0 YYMM 95-101;
run;
.
.
.
data CRSP1612;
infile " C:\QCF 2Ed\Data092018\SAS\Data\CRSP\CRSP1612.txt" lrecl=102;
input permno 1-6 cusip $ 8-15 @17 mcap f9.2 @27 ret f7.4 @35 vwretd f7.4 @43
sprtrn f7.4 @85 VOL f8.0 YYMM 95-101;
run;
data CRSPBeta;
set CRSP1201;
proc append base=CRSPBeta data=CRSP1202;
proc append base=CRSPBeta data=CRSP1203;
.
.
.
proc append base=CRSPBeta data=CRSP1611;
proc append base=CRSPBeta data=CRSP1612;
run;
data IBM;
set CRSPBeta;
if Cusip ne '45920010' then delete;
IBMRet=ret;
proc print data=IBM;
 var YYMM ret sprtrn vwretd;
run;
Title OLS SP IBM Beta;
proc reg;
model ret=sprtrn;
run;
```

Output 3.2a: Regression Result of IBM Against S&P 500

The REG Procedure
Model: MODEL1
Dependent Variable: ret

Number of Observations Read	60
Number of Observations Used	60

Analysis of Variance

Source	DF	Sum of Squares	Mean Square	F Value	Pr > F
Model	1	0.04883	0.04883	28.25	<.0001
Error	58	0.10027	0.00173		
Corrected Total	59	0.14910			

Root MSE	0.04158	**R-Square**	0.3275
Dependent Mean	0.00168	**Adj R-Sq**	0.3159
Coeff Var	2477.34120		

Parameter Estimates

Variable	DF	Parameter Estimate	Standard Error	t Value	Pr > |t|
Intercept	1	-0.00805	0.00567	-1.42	0.1612
Sprtrn	1	0.96320	0.18123	5.31	<.0001

The estimated IBM beta is 0.963. IBM is a defensive stock because its beta is less than one. The beta's t-statistic is 5.31, highly statistically significant, but that only means that its beta is greater than zero. The IBM t-statistic is -.204 [(0.963 – 1.0)/.181], when its beta is tested for statistic difference from one, the market average. Thus, IBM moves with the S&P 500.

The studentized residuals, RStudent exceeding 2.0 indicate the traditional presence of outliers. When a scaled residual known as the Cov Ratio distance measure exceeds 1+ 3p/n (or 1+[3*1/60 = 1.05]), it identifies influential observations that might be confirmed as outliers. The influential observations and the traditional outlier observations are denoted in BOLD.

Output 3.2b: IBM Beta Estimation Output Statistics

Obs	Residual	RStudent	Hat Diag H	Cov Ratio	DFFITS	DFBETAS	
						Intercept	sprtrn
1	0.0135	0.3273	0.0380	**1.0722**	0.0650	0.0251	0.0487
2	-0.005758	-0.1397	0.0343	**1.0715**	-0.0263	-0.0113	-0.0189
3	0.0385	0.9369	0.0252	1.0302	0.1507	0.0877	0.0877
4	0.007772	0.1875	0.0226	**1.0580**	0.0285	0.0279	-0.0145
5	0.004041	0.1026	0.1174	**1.1726**	0.0374	0.0245	-0.0346

Obs	Residual	RStudent	Hat Diag H	Cov Ratio	DFFITS	DFBETAS	
						Intercept	sprtrn
6	-0.0162	-0.3932	0.0332	**1.0652**	-0.0729	-0.0323	-0.0514
7	-0.002088	-0.0502	0.0168	**1.0530**	-0.0066	-0.0060	-0.0006
8	-0.0124	-0.2992	0.0185	**1.0516**	-0.0410	-0.0328	-0.0128
9	0.0494	1.2061	0.0204	1.0051	0.1742	0.1247	0.0749
10	-0.0352	-0.8588	0.0336	1.0443	-0.1603	-0.1435	0.1138
11	-0.0132	-0.3190	0.0177	**1.0503**	-0.0428	-0.0426	0.0102
12	0.009010	0.2167	0.0168	**1.0514**	0.0284	0.0276	-0.0029
13	0.0196	0.4799	0.0475	**1.0783**	0.1072	0.0322	0.0864
14	-0.009443	-0.2272	0.0167	**1.0511**	-0.0296	-0.0277	-0.0010
15	0.0355	0.8641	0.0294	1.0394	0.1504	0.0752	0.0990
16	-0.0598	-1.4652	0.0179	0.9791	-0.1977	-0.1640	-0.0516
17	0.0197	0.4755	0.0188	1.0470	0.0659	0.0514	0.0224
18	-0.0588	-1.4485	0.0286	0.9916	-0.2487	-0.2315	0.1608
19	-0.0190	-0.4655	0.0462	**1.0773**	-0.1024	-0.0318	-0.0819
20	-0.0224	-0.5493	0.0492	**1.0776**	-0.1250	-0.1016	0.1016
21	-0.004559	-0.1100	0.0240	**1.0604**	-0.0172	-0.0105	-0.0095
22	-0.0671	-1.6720	0.0393	0.9793	-0.3381	-0.1257	-0.2565
23	-0.0110	-0.2660	0.0228	**1.0569**	-0.0406	-0.0261	-0.0210
24	0.0292	0.7068	0.0201	1.0384	0.1013	0.0737	0.0420
25	-0.0158	-0.3874	0.0563	**1.0915**	-0.0947	-0.0744	0.0794
26	0.0199	0.4854	0.0374	**1.0667**	0.0956	0.0375	0.0712
27	0.0409	0.9920	0.0169	1.0177	0.1299	0.1268	-0.0139
28	0.0228	0.5492	0.0170	1.0422	0.0721	0.0707	-0.0094
29	-0.0682	-1.6814	0.0189	0.9580	-0.2335	-0.1814	-0.0807
30	-0.0271	-0.6557	0.0182	1.0389	-0.0893	-0.0725	-0.0260
31	0.0800	**2.0022**	0.0287	**0.9305**	0.3443	0.3202	-0.2231
32	-0.0193	-0.4675	0.0311	1.0605	-0.0838	-0.0396	-0.0571
33	0.0102	0.2464	0.0291	1.0642	0.0427	0.0396	-0.0279
34	-0.1483	**-4.0539**	0.0199	**0.6365**	-0.5781	-0.4249	-0.2338
35	-0.0225	-0.5423	0.0206	1.0464	-0.0787	-0.0559	-0.0344
36	0.001394	0.0336	0.0206	**1.0571**	0.0049	0.0048	-0.0021
37	-0.006492	-0.1588	0.0488	**1.0875**	-0.0359	-0.0293	0.0292
38	0.0187	0.4587	0.0548	**1.0874**	0.1104	0.0279	0.0921
39	0.0159	0.3858	0.0310	**1.0630**	0.0690	0.0631	-0.0470

Obs	**Residual**	**RStudent**	**Hat Diag H**	**Cov Ratio**	**DFFITS**	**DFBETAS**	
						Intercept	**sprtrn**
40	0.0671	1.6506	0.0167	0.9591	0.2152	0.2071	-0.0116
41	-0.004065	-0.0978	0.0167	**1.0526**	-0.0127	-0.0120	-0.0002
42	-0.0129	-0.3140	0.0350	**1.0693**	-0.0598	-0.0530	0.0433
43	-0.0150	-0.3620	0.0184	**1.0500**	-0.0496	-0.0397	-0.0153
44	-0.0107	-0.2705	0.1171	**1.1697**	-0.0985	-0.0646	0.0912
45	0.0138	0.3359	0.0420	**1.0765**	0.0703	0.0596	-0.0546
46	-0.1056	**-2.8671**	0.1176	**0.8963**	-1.0469	-0.0600	-0.9699
47	0.0122	0.2930	0.0184	**1.0517**	0.0401	0.0401	-0.0124
48	0.0120	0.2910	0.0311	**1.0655**	0.0522	0.0476	-0.0356
49	-0.0363	-0.9128	0.0869	**1.1015**	-0.2816	-0.1984	0.2531
50	0.0724	1.7926	0.0205	**0.9472**	0.2593	0.2575	-0.1121
51	0.1003	**2.6344**	0.0760	**0.8905**	0.7557	0.1194	0.6678
52	-0.0310	-0.7483	0.0177	1.0337	-0.1005	-0.1001	0.0243
53	0.0563	1.3766	0.0172	0.9868	0.1820	0.1595	0.0315
54	-0.005519	-0.1328	0.0183	**1.0540**	-0.0181	-0.0181	0.0054
55	0.0320	0.7774	0.0290	1.0441	0.1344	0.0681	0.0877
56	0.007104	0.1711	0.0191	**1.0545**	0.0239	0.0239	-0.0085
57	0.009004	0.2169	0.0191	**1.0538**	0.0303	0.0302	-0.0108
58	-0.005766	-0.1398	0.0332	**1.0702**	-0.0259	-0.0233	0.0183
59	0.0397	0.9680	0.0277	1.0307	0.1634	0.0867	0.1031
60	0.0137	0.3304	0.0179	**1.0503**	0.0446	0.0369	0.0118

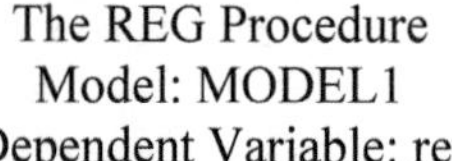

The REG Procedure
Model: MODEL1
Dependent Variable: ret

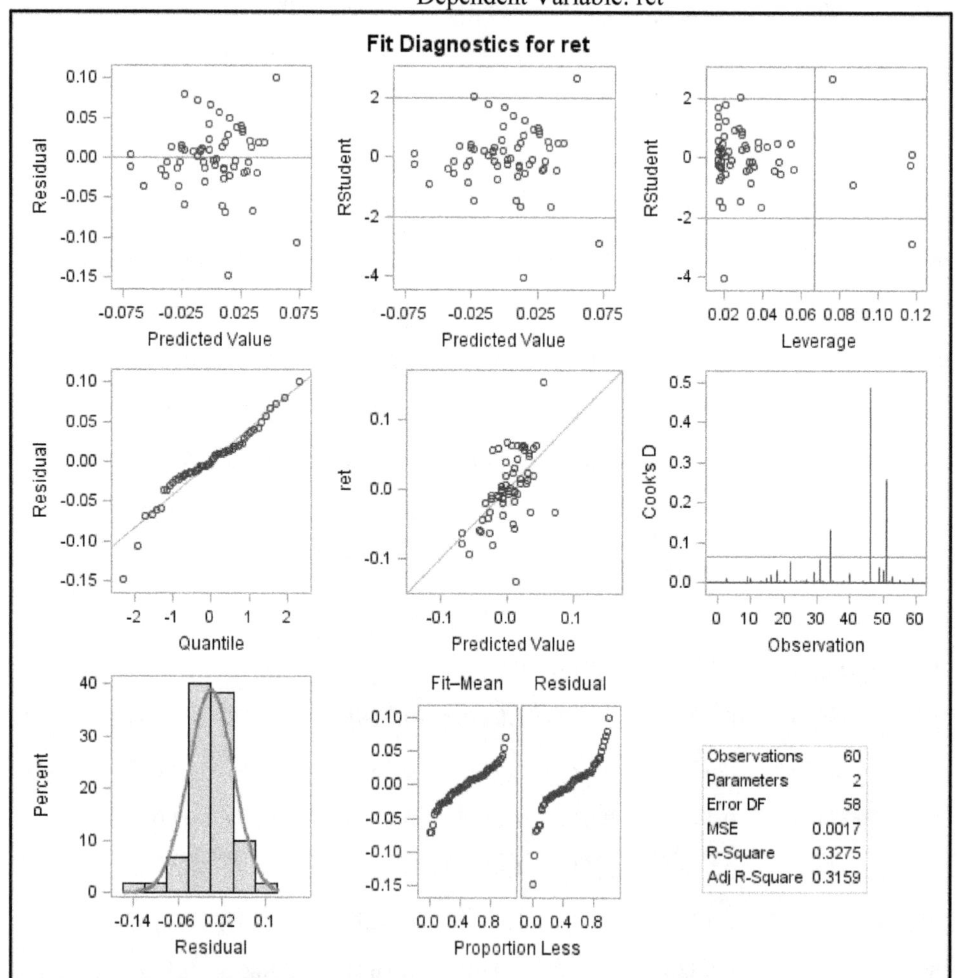

One sees from the presence of several (four) large studentized residuals, those exceeding 2.0 and less than -2.0, that the application of robust regression using PROC ROBUSTREG is appropriate. Robust regression is introduced in the Appendix to this chapter, and more completely addressed in Chapter 4.

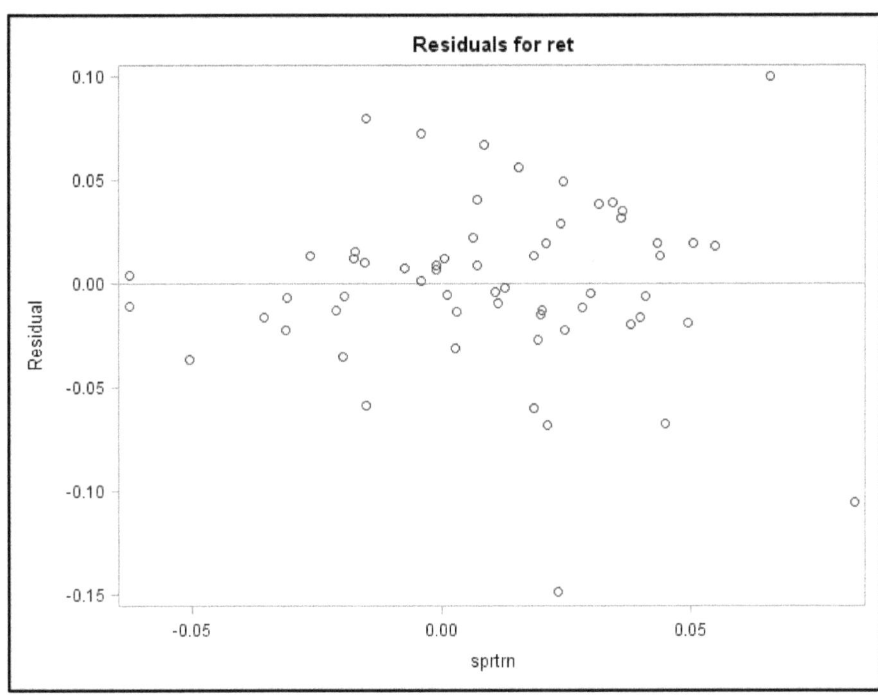

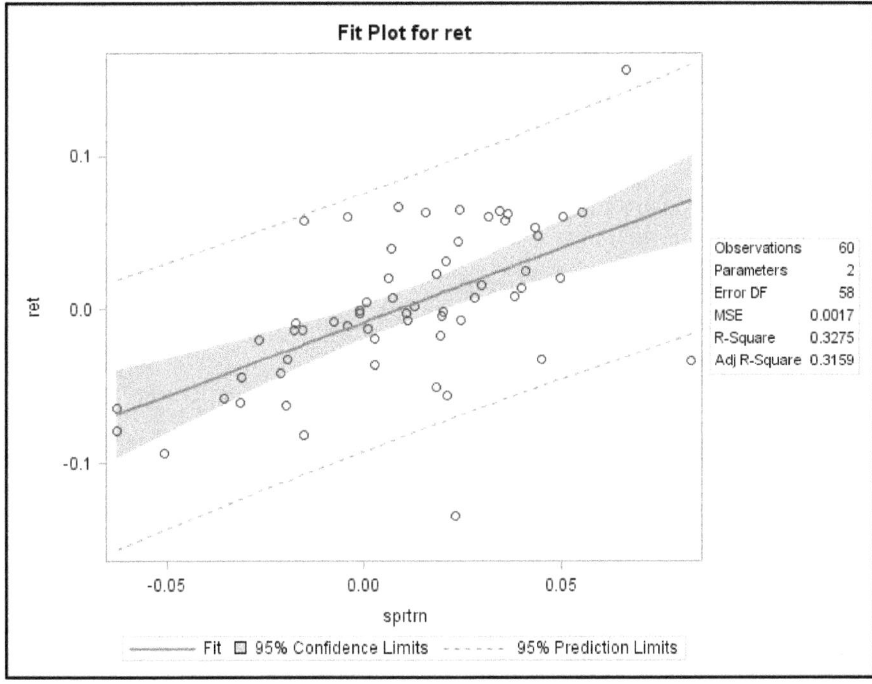

The IBM beta regression appears to have outlier issues versus the S&P index. Are outlier issues unique to regression against S&P index?

Program 3.3: IBM Beta Estimation Against CRSP Index

```
Proc reg data= IBM;
model ret=vwretd;
run;
```

Output 3.3: Regression Result of IBM Against CRSP Index

The REG Procedure
Model: MODEL1
Dependent Variable: ret

Number of Observations Read	60
Number of Observations Used	60

Analysis of Variance

Source	DF	Sum of Squares	Mean Square	F Value	Pr > F
Model	1	0.05079	0.05079	29.97	<.0001
Error	58	0.09831	0.00169		
Corrected Total	59	0.14910			

Root MSE	0.04117	**R-Square**	0.3407
Dependent Mean	0.00168	**Adj R-Sq**	0.3293
Coeff Var	2453.00675		

Parameter Estimates

Variable	DF	Parameter Estimate	Standard Error	t Value	Pr > |t|
Intercept	1	-0.00870	0.00564	-1.54	0.1286
vwretd	1	0.96540	0.17635	5.47	<.0001

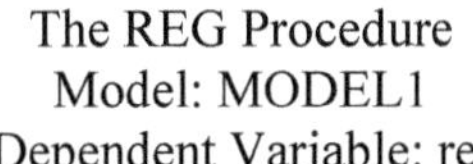

The REG Procedure
Model: MODEL1
Dependent Variable: ret

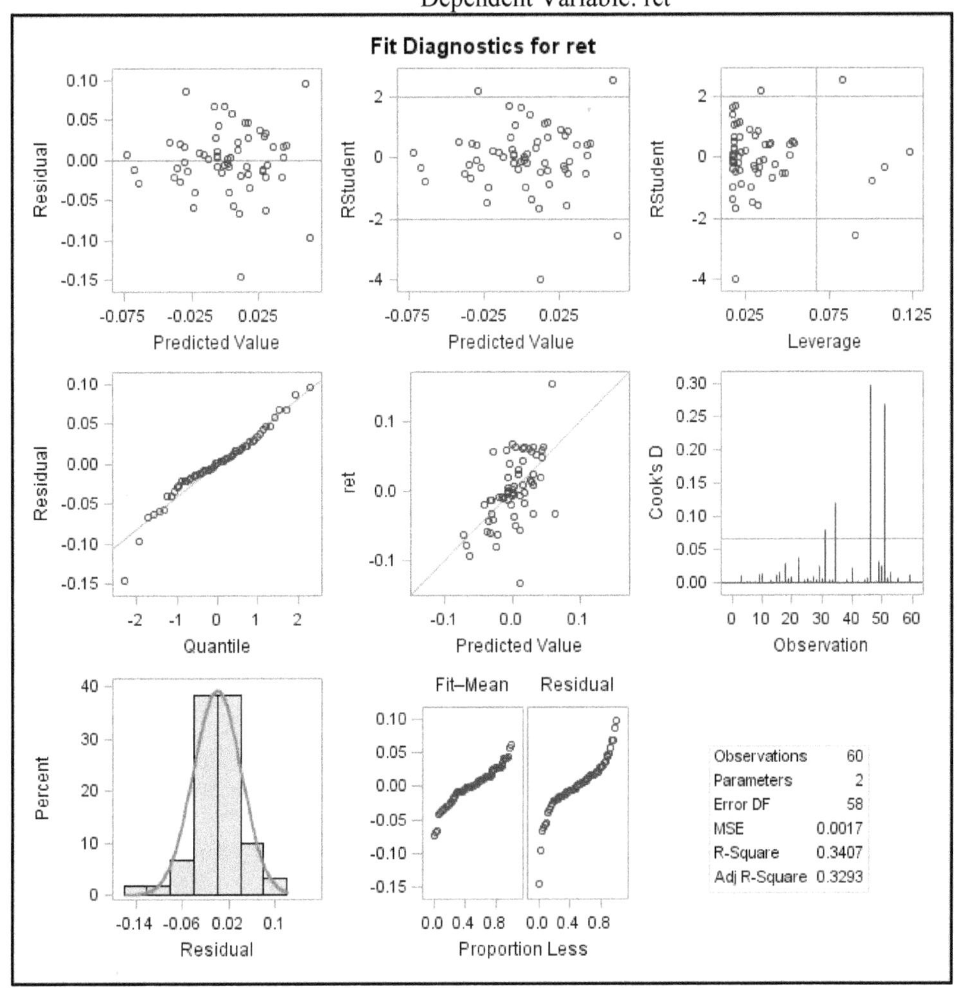

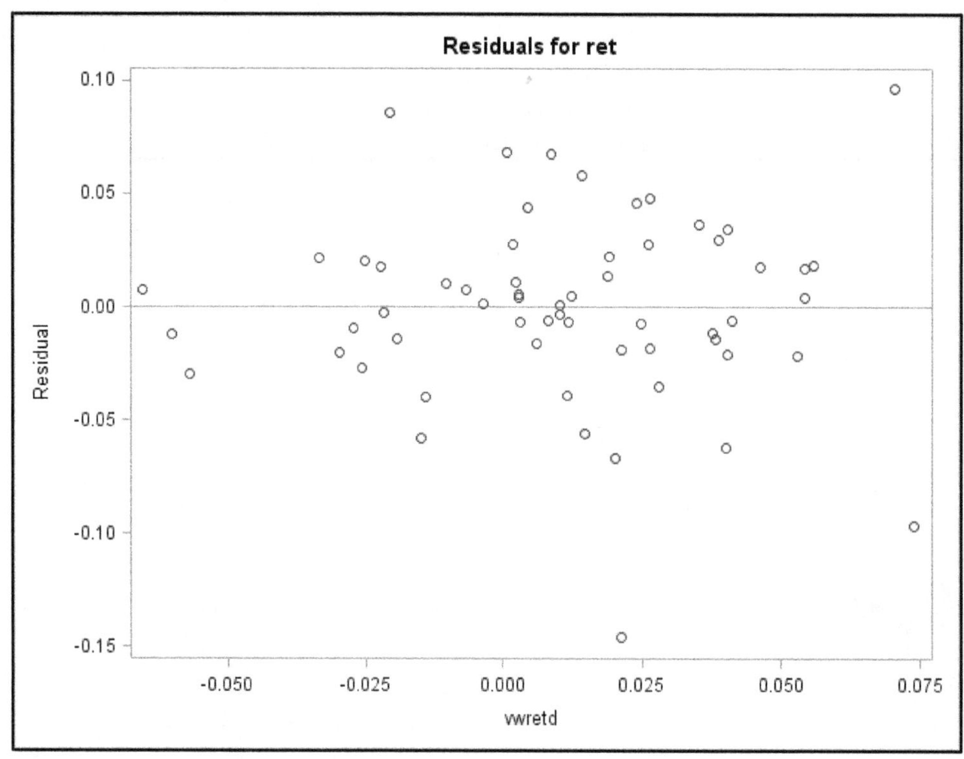

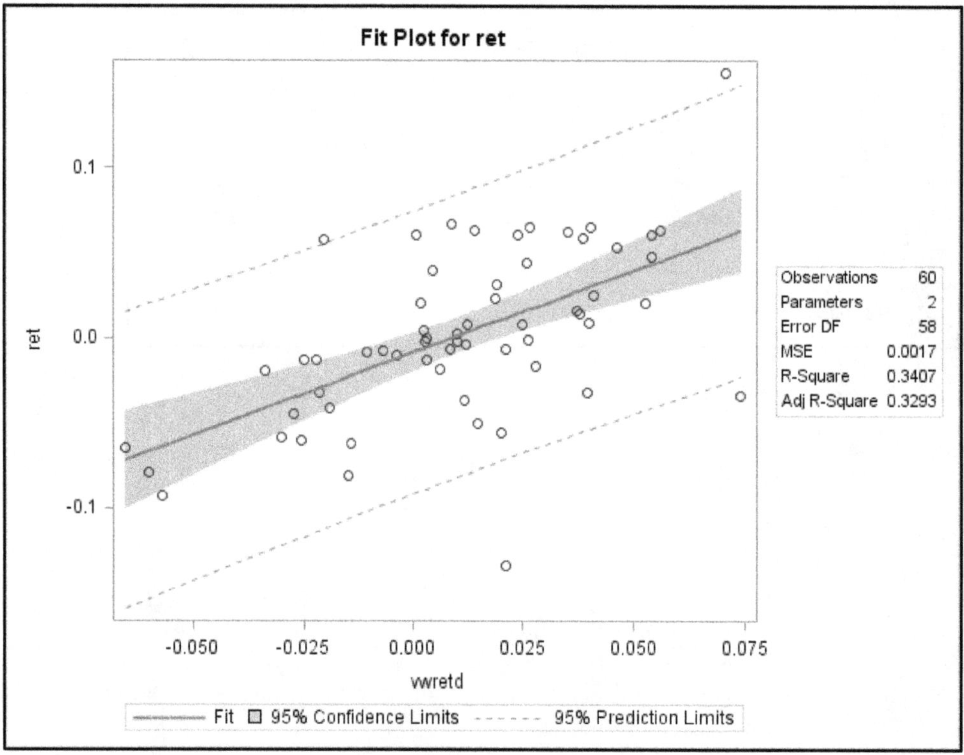

The IBM beta regression appears to have outlier issues versus the value-weighted CRSP Index, as well as the S&P 500 index. The outlier issues justify the use of robust regression.

Program 3.4: Robust Estimation of IBM Beta Against S&P 500 Index

```
proc robustreg data= IBM method=MM (CHIF=Tukey);
model ret=sprtrn;
run;
```

Output 3.4: Robust Regression Result of IBM Against S&P 500

The ROBUSTREG Procedure

Model Information	
Data Set	WORK.IBM
Dependent Variable	ret
Number of Independent Variables	1
Number of Observations	60
Method	MM Estimation

Profile for the Initial LTS Estimate	
Total Number of Observations	60
Number of Squares Minimized	45
Number of Coefficients	2
Highest Possible Breakdown Value	0.2667

MM Profile	
Chi Function	Tukey
K1	7.0410
Efficiency	0.9900

Parameter Estimates

Parameter	DF	Estimate	Standard Error	95% Confidence Limits		Chi-Square	Pr > ChiSq
Intercept	1	-0.0071	0.0050	-0.0170	0.0028	1.99	0.1582
sprtrn	1	1.0289	0.1641	0.7072	1.3505	39.31	<.0001
Scale	0	0.0351					

Diagnostics

Obs	Mahalanobis Distance	Robust MCD Distance	Leverage	Standardized Robust Residual	Outlier
5	2.4373	2.6446	*	0.2058	
34	0.4386	0.2895		-4.3013	*
44	2.4339	2.6412	*	-0.2137	
46	2.4408	2.3321	*	-3.1951	*

Diagnostics Summary

Observation Type	Proportion	Cutoff
Outlier	0.0333	3.0000
Leverage	0.0500	2.2414

The robust regression-estimated SAS beta for IBM is 1.029, higher than the S&P 500 Index, for 2012–2016, having identified two outliers.

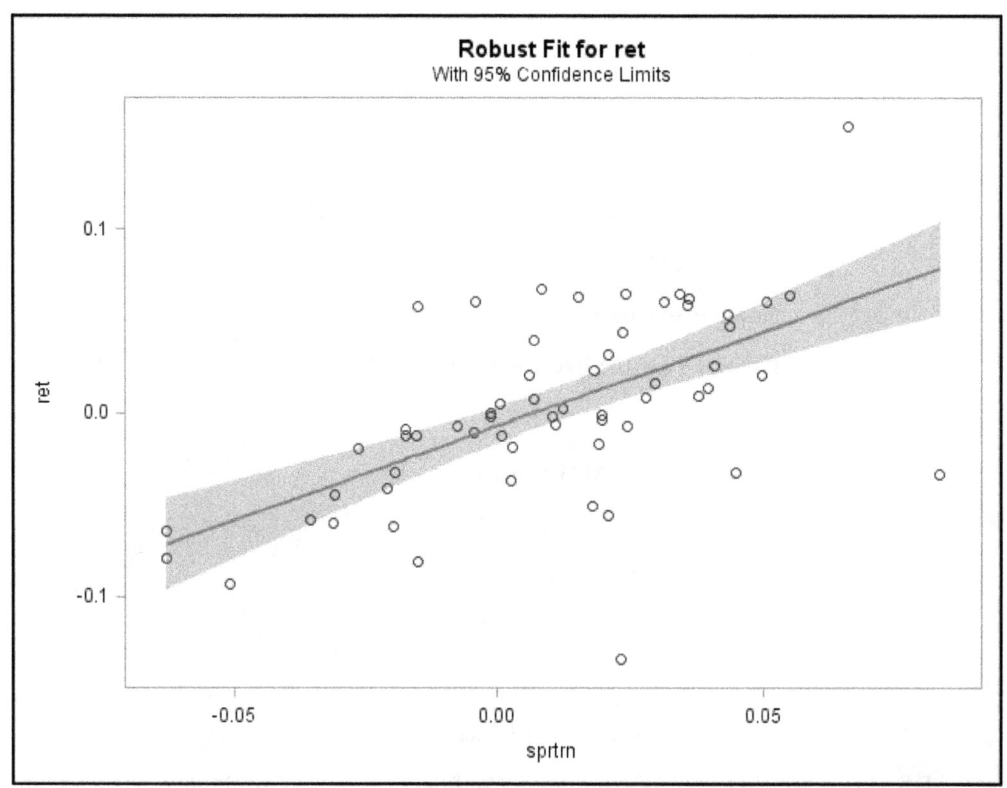

Goodness-of-Fit

Statistic	Value
R-Square	0.3302
AICR	74.0935
BICR	78.5812
Deviance	0.0865

Program 3.5: Robust Estimation of IBM Beta Against CRSP Index

```
proc robustreg data= IBM method=MM (CHIF=Tukey);
model ret=vwretd;
run;
```

Output 3.5: Robust Regression Result of IBM Against CRSP

The ROBUSTREG Procedure

Model Information	
Data Set	WORK.IBM
Dependent Variable	ret
Number of Independent Variables	1
Number of Observations	60
Method	MM Estimation

Profile for the Initial LTS Estimate	
Total Number of Observations	60
Number of Squares Minimized	45
Number of Coefficients	2
Highest Possible Breakdown Value	0.2667

MM Profile	
Chi Function	Tukey
K1	7.0410
Efficiency	0.9900

Parameter Estimates

Parameter	DF	Estimate	Standard Error	95% Confidence Limits		Chi-Square	Pr > ChiSq
Intercept	1	-0.0079	0.0052	-0.0180	0.0022	2.36	0.1247
vwretd	1	1.0008	0.1627	0.6820	1.3196	37.86	<.0001
Scale	0	0.0361					

Diagnostics

Obs	Mahalanobis Distance	Robust MCD Distance	Leverage	Standardized Robust Residual	Outlier
5	2.5088	2.6700	*	0.2512	

Diagnostics					
Obs	**Mahalanobis Distance**	**Robust MCD Distance**	**Leverage**	**Standardized Robust Residual**	**Outlier**
34	0.3439	0.2417		-4.0841	*
44	2.3278	2.4853	*	-0.3062	
49	2.2291	2.3845	*	-0.7831	

Diagnostics Summary		
Observation Type	**Proportion**	**Cutoff**
Outlier	0.0167	3.0000
Leverage	0.0500	2.2414

The robust regression-estimated SAS beta for IBM is 1.029, higher than the value-weighted CRSP Index, for 2012–2016, having identified one outlier.

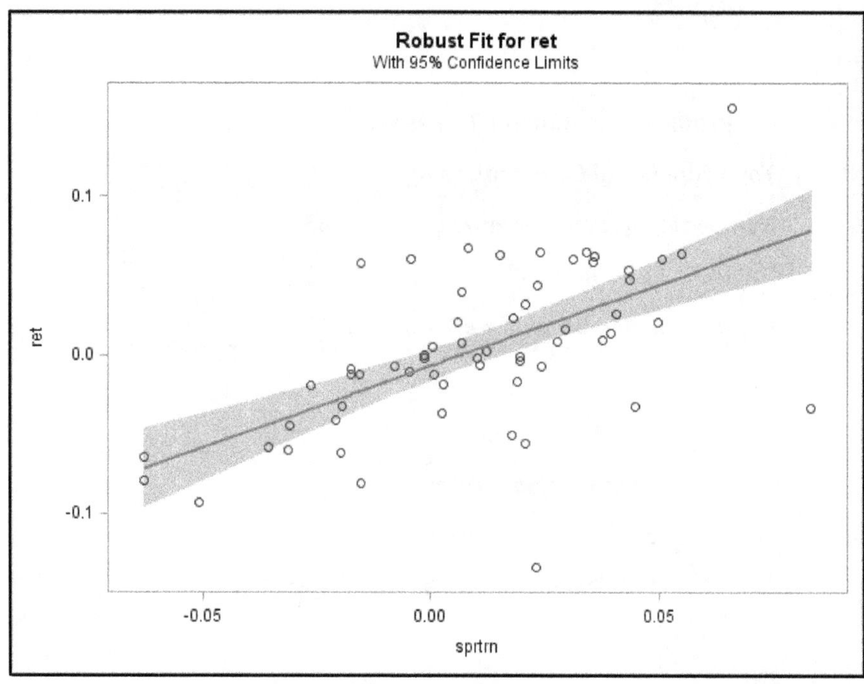

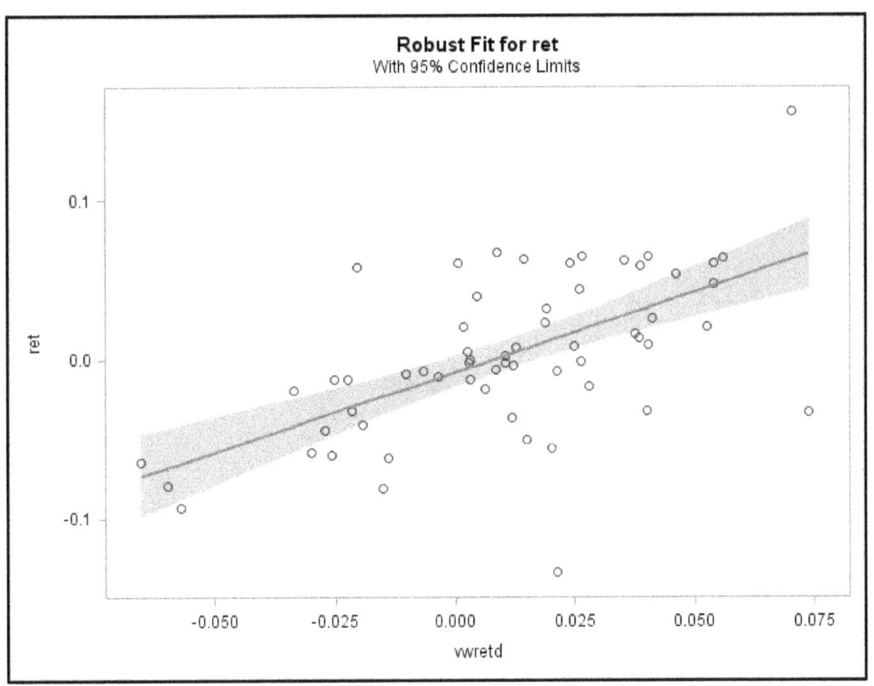

Goodness-of-Fit	
Statistic	**Value**
R-Square	0.3354
AICR	70.1307
BICR	74.6682
Deviance	0.0865

We have estimated OLS and Robust egression IBM betas. Outliers have been identified and more statistically efficient parameters estimated.

Program 3.6: Estimation of DuPont Beta Against S&P 500 Index

```
proc reg data= DD;
model ret=sprtrn;
run;
```

Output 3.6: Regression Result of DD Against S&P 500

The REG Procedure
Model: MODEL1
Dependent Variable: ret

Number of Observations Read	60
Number of Observations Used	60

Analysis of Variance

Source	DF	Sum of Squares	Mean Square	F Value	Pr > F
Model	1	0.15384	0.15384	50.44	<.0001

Analysis of Variance					
Source	**DF**	**Sum of Squares**	**Mean Square**	**F Value**	**Pr > F**
Error	58	0.17691	0.00305		
Corrected Total	59	0.33075			

Root MSE	0.05523	**R-Square**	0.4651
Dependent Mean	0.01386	**Adj R-Sq**	0.4559
Coeff Var	398.42552		

Parameter Estimates					
Variable	**DF**	**Parameter Estimate**	**Standard Error**	**t Value**	**Pr > |t|**
Intercept	1	-0.00340	0.00753	-0.45	0.6532
sprtrn	1	1.70960	0.24073	7.10	<.0001

The estimated DD beta is 1.710. DD is an offensive stock because its beta is greater than one. The beta's t-statistic is 7.10, highly statistically significant, but that only means that its beta is greater than zero. The DD t-statistic is 2.95 [(1.71 – 1.0)/.241] when its beta is tested for statistic difference from one, the market average. Thus, DD moves much greater than the S&P 500 Index.

Program 3.7: Robust Estimation of DD Beta Against S&P 500 Index

```
proc robustreg data= DD method=MM (CHIF=Tukey);
model ret=sprtrn;
run;
```

Output 3.7: Robust Regression Result of DD against S&P 500

The ROBUSTREG Procedure

Model Information	
Data Set	WORK.DD
Dependent Variable	ret
Number of Independent Variables	1
Number of Observations	60
Method	MM Estimation

Profile for the Initial LTS Estimate	
Total Number of Observations	60
Number of Squares Minimized	45
Number of Coefficients	2
Highest Possible Breakdown Value	0.2667

MM Profile	
Chi Function	Tukey
K1	7.0410
Efficiency	0.9900

Parameter Estimates							
Parameter	**DF**	**Estimate**	**Standard Error**	**95% Confidence Limits**		**Chi-Square**	**Pr > ChiSq**
Intercept	1	-0.0045	0.0067	-0.0176	0.0087	0.44	0.5050
sprtrn	1	1.5614	0.2196	1.1311	1.9917	50.58	<.0001
Scale	0	0.0459					

Diagnostics					
Obs	**Mahalanobis Distance**	**Robust MCD Distance**	**Leverage**	**Standardized Robust Residual**	**Outlier**
5	2.4373	2.6446	*	0.2868	
44	2.4339	2.6412	*	0.7104	
46	2.4408	2.3321	*	4.1452	*
50	0.4754	0.6430		3.7421	*

Diagnostics Summary		
Observation Type	**Proportion**	**Cutoff**
Outlier	0.0333	3.0000
Leverage	0.0500	2.2414

The robust regression-estimated SAS beta for DD is 1.561, much higher than the S&P 500 Index, for 2012–2016, having identified two outliers.

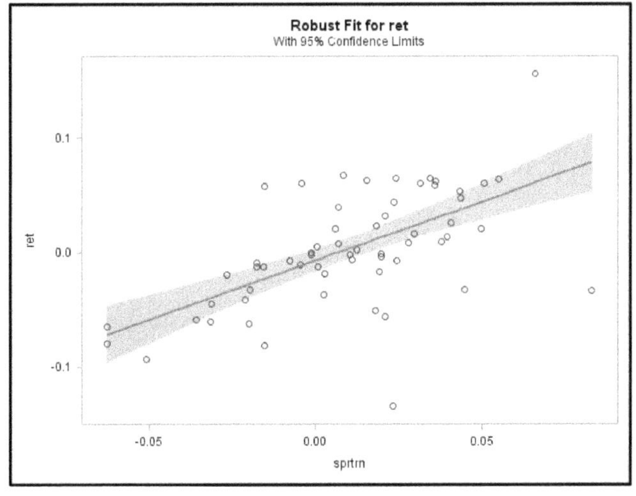

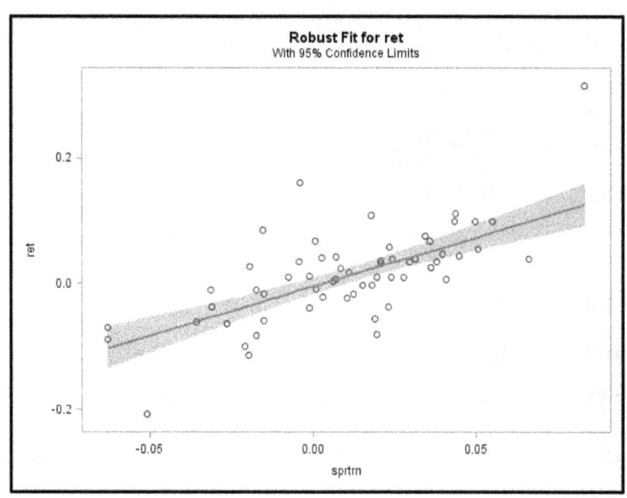

Goodness-of-Fit	
Statistic	**Value**
R-Square	0.3612
AICR	76.0392
BICR	80.4903
Deviance	0.1523

Program 3.8: Estimation of Boeing Beta Against S&P 500 Index

```
proc reg data= BA;
model ret=sprtrn;
run;
```

Output 3.8: Regression Result of BA Against S&P 500

The REG Procedure
Model: MODEL1
Dependent Variable: ret

Number of Observations Read	60
Number of Observations Used	60

Analysis of Variance

Source	DF	Sum of Squares	Mean Square	F Value	Pr > F
Model	1	0.05845	0.05845	24.88	<.0001
Error	58	0.13627	0.00235		
Corrected Total	59	0.19473			

Root MSE	0.04847	**R-Square**	0.3002
Dependent Mean	0.01636	**Adj R-Sq**	0.2881
Coeff Var	296.34434		

Parameter Estimates

Variable	DF	Parameter Estimate	Standard Error	t Value	Pr > |t|
Intercept	**1**	0.00571	0.00661	0.86	0.3909
sprtrn	**1**	1.05381	0.21128	4.99	<.0001

The estimated BA beta is 1.054. BA is an offensive stock because its beta is greater than one. The beta's t-statistic is 4.99, highly statistically significant, but that only means that its beta is greater than zero. The BA t-statistic is 0.46 when its beta is tested for statistic difference from one, and is not statistically different the market average. Thus, BA moves (slightly) greater the S&P 500 Index for 2012-2016.

The REG Procedure
Model: MODEL1
Dependent Variable: ret

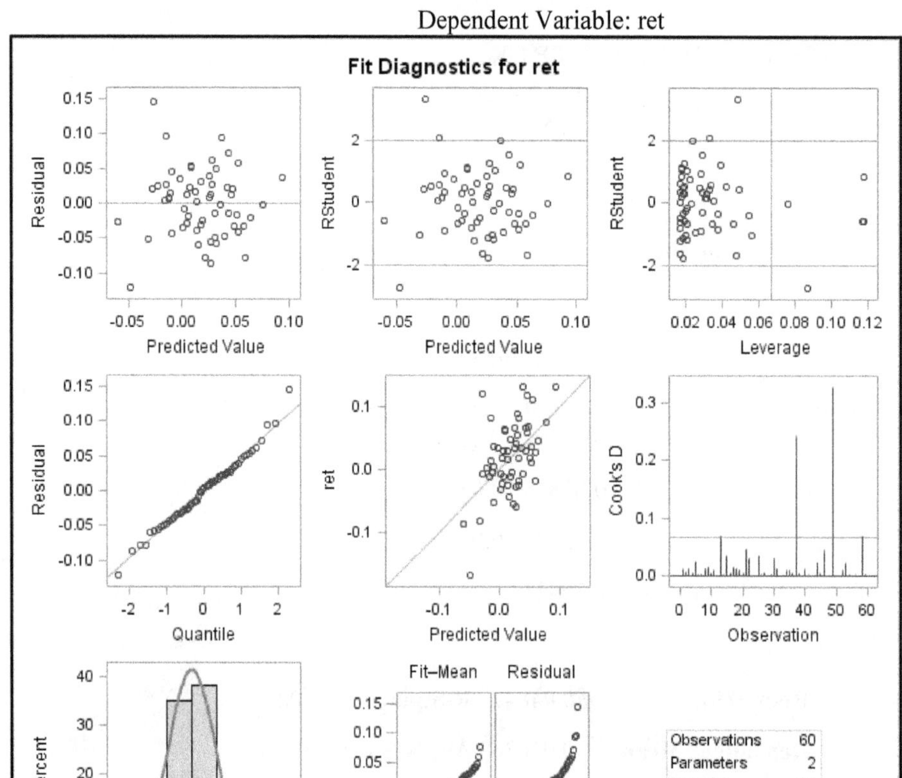

Program 3.9: Robust Estimation of Boeing Beta against S&P 500 Index

```
proc robustreg data= BA method=MM (CHIF=Tukey);
model ret=sprtrn;
run;
```

Output 3.9: Robust Regression Result of BA Against S&P 500

The ROBUSTREG Procedure

Model Information	
Data Set	WORK.BA
Dependent Variable	ret
Number of Independent Variables	1
Number of Observations	60
Method	MM Estimation

Profile for the Initial LTS Estimate	
Total Number of Observations	60
Number of Squares Minimized	45
Number of Coefficients	2
Highest Possible Breakdown Value	0.2667

MM Profile	
Chi Function	Tukey
K1	7.0410
Efficiency	0.9900

Parameter Estimates

Parameter	DF	Estimate	Standard Error	95% Confidence Limits		Chi-Square	Pr > ChiSq
Intercept	1	0.0051	0.0065	-0.0077	0.0179	0.60	0.4378
sprtrn	1	1.0667	0.2092	0.6567	1.4768	26.00	<.0001
Scale	0	0.0472					

Diagnostics

Obs	Mahalanobis Distance	Robust MCD Distance	Leverage	Standardized Robust Residual	Outlier
5	2.4373	2.6446	*	-0.5529	
37	1.3760	1.5618		3.1027	*
44	2.4339	2.6412	*	-0.5424	
46	2.4408	2.3321	*	0.7861	

Diagnostics Summary

Observation Type	Proportion	Cutoff
Outlier	0.0167	3.0000
Leverage	0.0500	2.2414

The robust regression-estimated SAS beta for BA is 1.067, higher than the S&P 500 Index for 2012–2016, having identified one outlier.

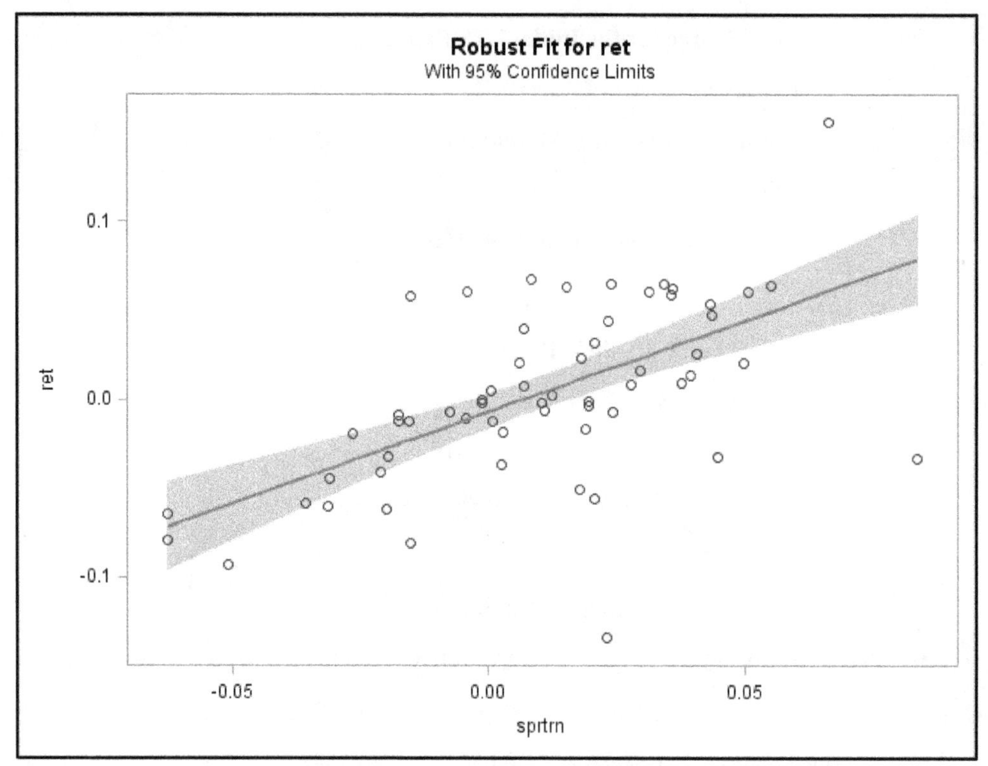

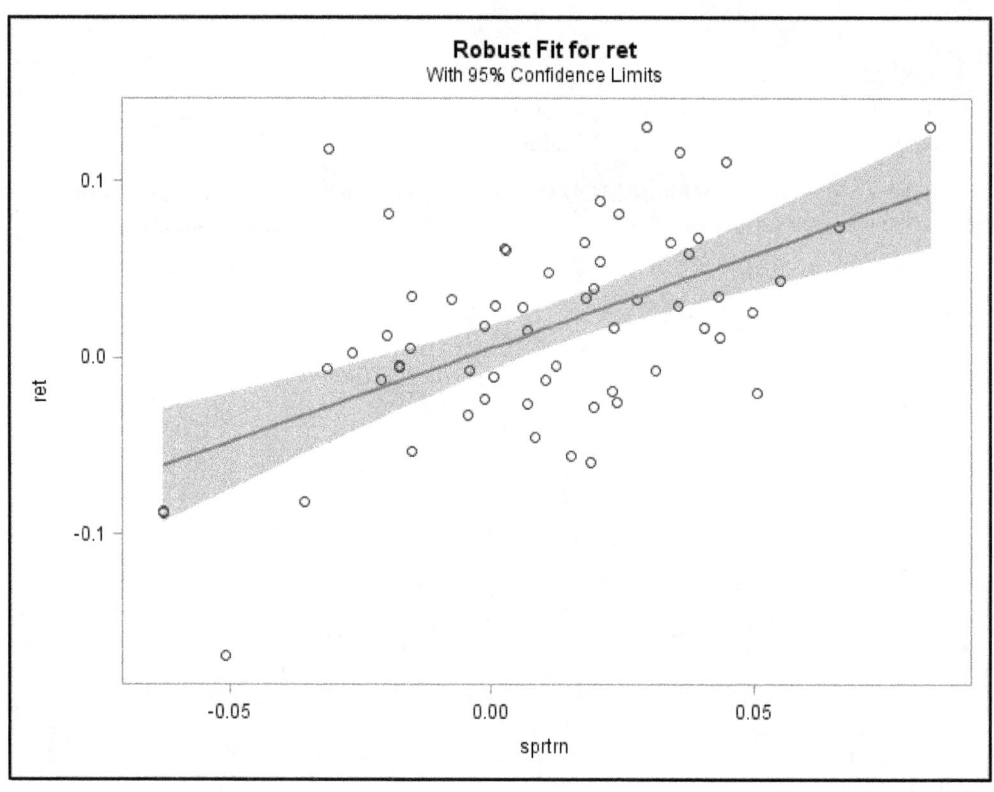

Goodness-of-Fit	
Statistic	**Value**
R-Square	0.2746
AICR	60.3788

Goodness-of-Fit	
Statistic	**Value**
BICR	65.0801
Deviance	0.1267

3.6 Multi-Beta Risk Models

The difficulty of measuring beta and its corresponding Security Market Line gave rise to extra-market measures of risk found in the work of King (1966), Farrell (1973), and other multi-beta risk models. Farrell (1973, 1997) estimated a four "factor" extra-market covariance model. Farrell took an initial universe of 100 stocks in 1973 (due to computer limitations), and ran market models to estimate betas and residuals from the market model:

$$R_{j_t} = a_j + b_j R_{M_t} + e_j \tag{23}$$

$$e_{j_t} = R_{j_t} - \widehat{a_j} - \widehat{b_j} R_{M_t} \tag{24}$$

The residuals of equation (24) should be independent variables. That is, after removing the market impact by estimating a beta, the residual of IBM should be independent of Dow, Merck, or Dominion Resources. The residuals should be independent, of course, in theory. Farrell (1973) examined the correlations among the security residuals of equation (24) and found that the residuals of IBM and Merck were highly correlated, but the residuals of IBM and D (then Virginia Electric & Power) were not correlated. Farrell used a statistical technique known as Cluster Analysis to create clusters, or groups, of securities, having highly correlated market model residuals. Farrell found four clusters of securities based on his extra-market covariance. The clusters contained securities with highly correlated residuals that were uncorrelated with residuals of securities in the other clusters. Farrell referred to his clusters as "Growth Stocks" (electronics, office equipment, drug, hospital supply firms, and firms with above-average earnings growth), "Cyclical Stocks" (metals, machinery, building supplies, general industrial firms, and other companies with above-average exposure to the business cycle), "Stable Stocks" (banks, utilities, retailers, and firms with below-average exposure to the business cycle), and "Energy Stocks" (coal, crude oil, and domestic and international oil firms).

In 1976, Ross published his "Arbitrage Theory of Capital Asset Pricing," which held that security returns were a function of several (4–5) economic factors. In 1986, Chen, Ross, and Roll developed and estimated a multi-factor security return model based on:

$$R = a + b_{MP}MP + b_{DEI}DEI + b_{UI}UI + b_{UPR}UPR + b_{UTS}UTS + e_t \tag{25}$$

where MP = monthly growth rate of industrial production,

DEI = change in expected inflation,

UI = unexpected inflation,

UPR = risk premium,

and UTS = term structure of interest rates.

Chen, Ross, and Roll (CRR) defined unexpected inflation as the monthly (first) differences of the Consumer Price Index (CIP) less the expected inflation rate. The risk premia variable is the "Baa and under" bond return at time and less the long-term government bond return. The term structure variable is the long-term government bond return less the Treasury bill rates, known at time t-1, and applied to time t. When CRR applied their five-factor model in conjunction with the value-weighted index betas during the 1958-1984 period, the index betas were not statistically significant whereas the economic variables were statistically significant. The Stone, Farrell, and Chen, Ross, and Roll multi-factor model used 4–5 factors to describe equity security risk. The models used different statistical approaches and economic models to control for risk. The reader might now ask a simple question; if four or five betas are not appropriate, why not estimate 20 betas? The important point is that the betas should be estimated on variables that are independent, or orthogonal, of the other variables. The estimation of 20 betas on orthogonal variables, 10–15 of which are not statistically significant, can produce quite similar expected returns for securities as the four or five betas estimated on economic variables, or pre-specified variables. Blin and Bender (1987–1997) estimated a 20 (factor) beta model of covariances based on 2.5 years of weekly stock returns data. The Blin and Bender Arbitrage Pricing Theory (APT) model followed the Ross factor modeling theory, but Blin and Bender estimated betas from 20 orthogonal factors. Blin and Bender never sought to identify their factors with economic variables. We can create 100 stock portfolios for the January 1993–December 2001 period using the consensus analysts' forecasts and revisions variable (CTEF), producing asset selection excess returns of 127 basis points (t=1.73), which is statistically significant at the 10 percent level.

The total excess return for a multiple factor model, referred to as the MFM, in the Rosenberg methodology for security j, at time t, dropping the subscript t for time, can be written:

$$E(R_j) = \sum_{k=1}^{K} \beta_{jk} \tilde{f}_k + \tilde{e}_j \tag{26}$$

The non-factor, or asset-specific, return on security j, is the residual risk of the security, after removing the estimated impacts of the K factors. The term, $\tilde{f}_k$, is the rate of return on factor k. A single factor model, in which the market return is the only estimated factor, is obviously the basis of the Capital Asset Pricing Model. We discuss multifactor risk models in the Chapter 5.

3.7 Summary and Conclusions

This chapter has introduced the reader to the risk-return trade-offs of Markowitz, Sharpe, Mossin, and Treynor. The Markowitz Efficient Frontier traces the trade-off of return and total risk, as measured by the portfolio standard deviation. The CAPM holds that the trade-off exists between return and systematic risk, as measured by the stock beta. Outliers are present in beta estimations and robust regression should be used.

3.8 Appendix: Robust Regression and SAS Implementation

In ordinary least squares analysis, the Population regression is:

$$y = x\beta + \epsilon \tag{A-1}$$

Where

$y = n \times 1$ column vector of the response (dependent variable)

$x = n \times p$ matrix of independent (explanatory) variables

$\beta: p \times 1$ column vector of regression parameters

$\varepsilon: n \times 1$ column vector of errors

$x(i): x$ matrix with i^{th} row deleted

Now let fitted line be

$$\hat{y} = xb \tag{A-2}$$

$$e = y - \hat{y} \tag{A-3}$$

Where

b: estimate of β

e: residual vector

s^2: estimated error variance

$b(i)$: estimate of β when i^{th} row is deleted

$s^2(i)$: estimated error variance when i^{th} row of x and y is deleted

Hat-Matrix Diagonals

The diagonal elements of the hat matrix are useful in detecting extreme points in the regression space where they tend to have larger values. The h_i are the diagonal elements of the least squares projection matrix, also known as the hat matrix.

$$H = x(x^T x)^{-1} x^T \tag{A-4}$$

$$\hat{y} = xb = Hy \tag{A-5}$$

$$h_i - \frac{1}{n} = \widetilde{h_i} = \tilde{x}_i(\tilde{x}^T \tilde{x})^{-1} \tilde{x}_i^T \tag{A-6}$$

where ~ denotes the data is centered (the mean is subtracted)

For the generalized linear model, the variance of the ith individual observation is

$$v_i = v(\pi_i) \tag{A-7}$$

For the ith observation, let

$$w_{ei} = v_i^{-1}\left(g'(\pi_i)\right)^{-2} \tag{A-8}$$

$g'(\pi_i)$ is the derivative of the link function evaluated at π_i. The weight matrix is a diagonal matrix with the ith diagonal element defined by (A-8), which are also used in computing the expected information matrix. Define the leverage, or hat-matrix diagonal, as the ith diagonal element of the matrix

$$w_e^{\frac{1}{2}} x(x_i w_e^i x)^{-1} x' w_e^{\frac{1}{2}} \tag{A-9}$$

If the estimated probability is extreme (less than 0.1 and greater than 0.9, approximately), then the hat-matrix diagonal might be greatly reduced in value. Consequently, when an observation has a very large or very small estimated probability, its leverage is not a good indicator of the observation's distance from the regression space.

Residuals

Residuals are useful in identifying observations that are not explained well by the model.

The Pearson residual is the square root of the ith observation's contribution to Pearson's chi-square:

$$r_{pi} = r_i \sqrt{\frac{1}{v(\pi_i)}} \tag{A-10}$$

The Pearson residual, standardized to have unit (asymptotic) variance is:

$$r_{spi} = \frac{r_{pi}}{\sqrt{1-h_i}} \tag{A-11}$$

The studentized residual, RStudent, is

$$e_i^* = \frac{e_i}{s(i)V_i - h_i} \tag{A-12}$$

The COVRATIO is

$$\frac{1}{\left[\frac{n-p-1}{n-p} + \frac{e_i^{*2}}{n-p}\right]^p (1-h_i)} \tag{A-13}$$

The cutoff COVRATIO is $1 - \frac{3p}{n}$

Points where the COVRATIO exceeds $3p/n$ should be investigated as possible outliers, or influential observations, see Belsley, Kuh, and Welsch, pp. 12–23

In robust regression, rather than throwing or removing observations whose residuals lie outside of a 95 percent confidence interval, observations are re-weighted so that essentially all observations are used and observation weights are inversely related to the OLS residuals; that is, the higher the OLS residual, the lower the weight of the observation in robust regression.

Residual Regression Plots

Regression plots of residuals with other columns are used to detect the presence of outlying rows or model specification errors. A plot of the residuals against the independent columns can detect whether the linear regression is appropriate and if the variance of the error terms is constant. When the function is thought to be nonlinear, or multiple regression has been used, a plot of the residual values against the predicted values is a means of determining the consistency of error variance.

Residual plots that involve independent columns cannot be selected if the independent columns have not been specified.

PROC ROBUSTREG Statement

The PROC ROBUSTREG statement invokes the ROBUSTREG procedure. Method option

```
METHOD=method-type <(options)>
```

specifies the estimation method and some additional options for the estimation method. PROC ROBUSTREG provides four estimation methods: M estimation, LTS estimation, S estimation, and MM estimation. The SAS default method is M estimation. We illustrate with M, S, LTS, and MM estimations in Chapter 4.

Two of the most widely used weighting functions for robust regression are the 1.345 Huber and the Beaton-Tukey bisquare of 4.6845.

Options with METHOD=LTS

When you specify METHOD=LTS<(options)>, you can specify the following options:

- `CSTEP=n`

  specifies the number of concentration steps (C-steps) for the LTS estimate;

- `IADJUST=ALL | NONE`

  requests (IADJUST=ALL) or suppresses (IADJUST=NONE) the intercept adjustment for all estimates in the LTS algorithm. By default, the intercept adjustment is used for data sets that contain fewer than 10,000 observations.

Options with METHOD=MM

PROC ROBUSTREG provides two MM functions, Tukey's bisquare function and Yohai's optimal function, which you can request by specifying CHIF=TUKEY and CHIF=YOHAI, respectively. The default is Tukey's bisquare function.

`EFF=value`

specifies the efficiency (as a fraction) of the MM estimate. The parameter in the function is determined by this efficiency. The default efficiency is set to 0.85, which corresponds to if you specify CHIF=TUKEY or if you specify CHIF=YOHAI.

A Brief History of Modern Robust Regression

With PROC ROBUSTREG you can use the Huber (1973) M estimation procedure, the Rousseeuw (1984) Least Trimmed Squares (LTS), the Rousseeuw and Yohai (1984) S procedure, or Yohai (1987) MM estimation procedure. We will report iterations of these procedures in Chapter 4 as we simulate various robust regression investment strategies to maximize portfolio returns. In this appendix, we will dive deeper into the Huber Maud MM procedure that we use on a daily basis for portfolio construction.

The Huber M estimation procedure does not maximize the sum of the squared errors, but rather the sum of the residuals as stated:

$$Q(\theta) = \sum_{i=1}^{n} \rho\left(\frac{r_i}{\sigma}\right) \tag{A-14}$$

where $r = y - x\theta$ and p is the quadratic function, Huber (1973, 1981) held that the robust procedure should be "optimal or nearly optimal." It should be robust in the sense that small deviations from the model assumptions only slightly impair the asymptotic variance of the estimate, and larger deviations from the model should not cause a "catastrophe" (Huber, 1981, p. 5). Huber was concerned with the efficiency of the parameter estimated. Robustness means insensitivity to small deviations from model assumptions and the minimizations off the degradation of performance for ε – deviations from the assumptions. Let T_n be an estimate

$$\sum \rho\,(x_i;\, T_n) = min! \tag{A-15}$$

or

$$\sum \psi\,(x_i;\, T_n) = 0 \tag{A-16}$$

where

$$\psi(x;\, \theta) = \frac{d}{d\theta}\rho(x;\theta) \tag{A-17}$$

In the case of linear fitting of (A-1), the first order conditions are

$$\sum_{i=1}^{n} \psi\left(\frac{r_i}{\sigma}\right) x_{ij} = 0\,, j = 1, \ldots, \rho \tag{A-18}$$

PROC ROBUSTREG solves (A-18) by iteratively reweighted least squares with the weight function

$$\mathrm{w}(x) \ = \frac{\psi(x)}{x} \tag{A-19}$$

The σ in (A-18) is unknown and must be estimated. Huber (1973) modifies the objection function (A-14) as

$$Q(\theta, \sigma) = \sum_{i=1}^{n} \left[\rho\left(\frac{r_i}{\sigma}\right) + a\right]\sigma \tag{A-20}$$

and $\hat{\sigma}$ is estimated by Huber (1973) as:

$$(\sigma^{m+1})^2 = \frac{1}{nh}\sum_{i=1}^{n} X_d\left(\frac{r_i}{\hat{\sigma}^{(m)}}\right)(\hat{\sigma}^{(m)})^2 \tag{A-21}$$

where

$$X_d(x) = \begin{cases} \frac{x^2}{2}, & |x| < d\ \frac{d^2}{2}, \\ 0, & otherwise. \end{cases} \tag{A-22}$$

An alternative to the Huber weighting function is the Beaton-Tukey (1974) bisquare function where σ is solved from:

$$\frac{1}{n-p}\sum X_d\left(\frac{r_i}{\sigma}\right) = \beta \tag{A-23}$$

with

$$X_d(x) = \frac{3x^2}{d^2} - \frac{3x^4}{d^4} + \frac{x^6}{d^6}$$

If $|x| < d$ otherwise $\beta = fX_d\, d\phi(s)$

What we need is a plain English explanation. The bisquare and Huber M estimation weight functions map the sensitivity of the robust estimator to the outlier value. The modeler can identify outliers, or influential data, and re-run the ordinary least squares regressions on the re-weighted data, a process referred to as robust (ROB) regression. In ordinary least squares, OLS, all data is equally weighted. The weights are 1.0. In robust regression one weights the data inversely with its OLS residual; i.e., the larger the residual, the smaller the weight of the observation in the robust regression. In robust regression, several weights may be used. We will review the Beaton-Tukey (1974) bisquare iteratively weighting scheme. The intuition is that the larger the estimated residual, the smaller than weight. The Beaton-Tukey bisquare, or biweight criteria, for re-weighting observations is:

$$w_i = \begin{cases} (1 - (\frac{|e_i|}{\sigma_\varepsilon}/4.685)^2)^2, & if\ \frac{|e_i|}{\sigma_\varepsilon} \geq 4.685, \\ 0, & if\ \frac{|e_i|}{\sigma_\varepsilon} < 4.685. \end{cases} \tag{A-24}$$

Doug Martin, co-author of a great book, *Robust Statistics: Theory and Methods (with R)*, suggested that the rho, psi, and weight graphs would be very helpful to the reader. The Huber weighting function (A-22) is depicted in Figure A.1. The Beaton-Tukey bisquare weighting function (A-24) can be depicted graphically in Figure A.2 (c).

Figure A.1: The Huber Function

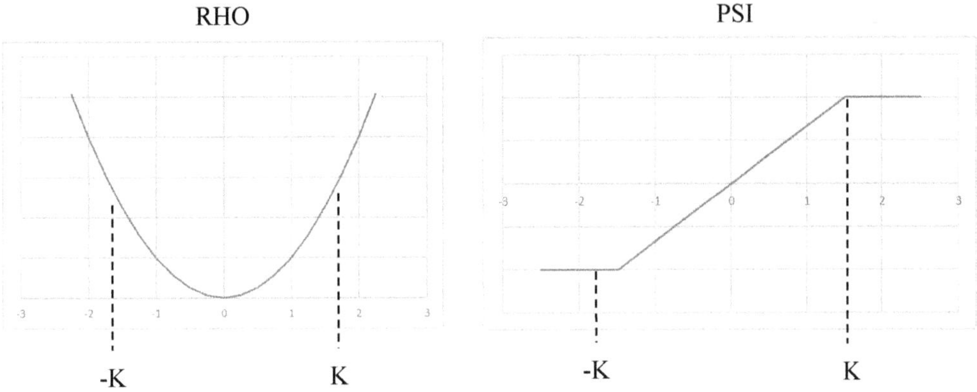

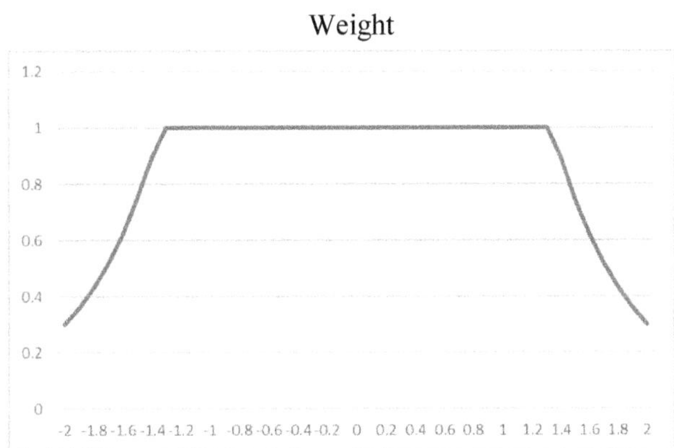

Figure A.2: The Tukey Bisquare Function

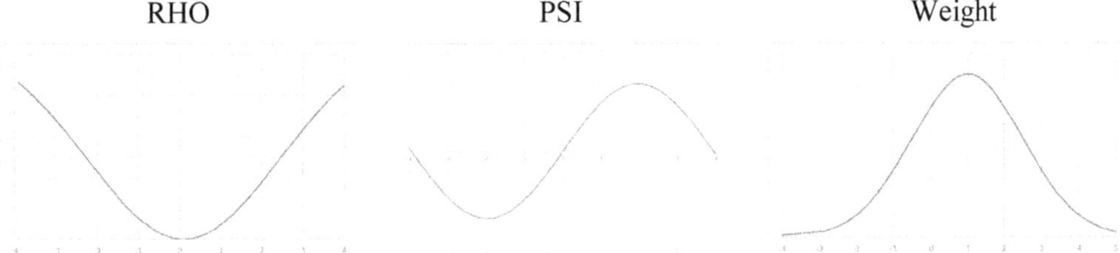

The Beaton-Tukey bisquare robust regression weighted scheme is one of eight schemes that were analyzed in Andrews, Bickel, Hampel, Huber, Rogers, and Tukey (1972). Most robust regression modeling techniques produce similar median forecasting errors, which are substantially lower than the OLS median forecasting errors.

[1] Bill Sharpe (1971) estimated robust betas using least absolute deviation regressions.

Chapter 4: Robust Regression and Stock Selection in Global Equity Markets

4.1 Introduction and Efficient Markets

In this chapter we build several models of stock selection. We apply many robust regression techniques to select stocks in various global stock universes. Robust regression applications are appropriate for modeling stock returns in global markets, producing highly statistically significant models.

In Chapter 2, you learned about income statements and balance sheets. Financial data was used to calculate ratios to provide objective criteria to assess the performance of firms. Financial data can be used to rank stocks on their attractiveness. We examined the credit worthiness and quality of firms using the Altman Z bankruptcy model. We proceeded to develop and estimate stock selection models to produce statistically significant outperformance.

Economists have long debated whether stock markets are efficient. In the 1960s and 1970s, most academicians argued that stock markets were efficient. That is, stock market prices reflected the information known at that point in time. Excess returns should not be earned. Roberts (1959) set forth three forms of stock market efficiency. In the weak form of the efficient markets hypothesis, stock prices reflect all information in the past price of stock and volume data. Economists should not regard technical analysis, the plotting and analysis of past stock prices, as a source of excess returns. Alexander (1961, 1964) and Fama and Blume (1966) tested filter rules such as: buy a stock if it falls y percent from its recent high; sell a stock and go short if it rises y percent above its recent low. Excess returns were found to be very small in these tests and completely eliminated by transactions costs. Technical analysis was thus useful.

In the second form of the efficient markets hypothesis – the semi-strong form – stock prices reflected all public information we learned about in Chapter 2, such as earnings, earnings announcements, book value, sales, and other measures of fundamental data. Fama, Fisher, Jensen, and Roll (FFJR, 1969) studied 940 stock splits from 1927 to 1959. Stocks split in times of rising prices, and fell into a "more favorable trading range." FFJR estimated the market model betas of Chapter 3 to account for market risk. Stocks that split increased in price for 2.5 years prior to the split announcement. However, after the split was announced, no additional excess returns were earned for 2.5 years following the split announcement. Fama (1970) cited the Ball and Brown (1968) study of 261 earnings announcements during the 1946–1966 time period that concluded that no more than 10–15 percent of the annual earnings announcement information had not been anticipated by the market as of the month of the announcement. Further, Waud (1971) found that Federal Reserve announcements were incorporated into stock prices. Thus, Fama (1970) concluded that markets were semi-strong efficient, which would doom fundamental analysis.

The third form of the efficient markets hypothesis – the strong form – held that stock prices reflected all information, including non-public information, such as insider trading and mutual fund performance. The Lorie and Niederhoffer (1968) and Jaffe (1974) studies of insider trading refuted the strong form efficiency studies. Insider trading yielded statistically significant excess returns. Mutual performance analysis of Jensen (1968) cast doubt on the excess returns earned by mutual fund professionals; they possessed no non-public knowledge. Thus, in 1970 and 1991, Fama stated that stocks markets were almost completely efficient and that prices reflected almost all information. An investor could not beat a buy and hold strategy.

Academicians cried "Efficient Markets!" Brealey (1969) and Stone (1970) published outstanding MIT Press monographs that reported and derived earnings random walks as market equilibrium and efficiency. Practitioners, often well-educated, and faculty members at Columbia Business School and UCLA, such as Graham and Dodd, cried out that academics were at least naïve if not completely wrong. Graham, and his Columbia student, Warren Buffet, managed funds that consistently beat the stock market, using the financial data discussed in Chapter 2. The Graham and Dodd classic *Security Analysis* (1934); Graham, Cottle, and Dodd *Security Analysis, Fourth Edition* (1962); and Graham, *The Intelligent Investor* (1950, 1975) presented fundamental analysis that had generated excess returns. Financial data, such as earnings, book value, cash flow, sales, forecasted earnings per share, have been used to generate excess returns in portfolios for many, many years. Applications of the low price-to-earnings ratio (low P/E) strategy been used by students of Benjamin Graham, such as Warren Buffet, since 1950. Academicians and practitioners have embraced the strategy since Basu (1977); Dreman (1979 and 1998); and Lakonishok, Shleifer, and Vishny (1994). The empirical evidence in these articles and books reports that the low P/E strategy is statistically significant at identifying mis-priced stocks. We agree! The low P/E strategy is relevant and statistically significant at an initial level. However, the application of various testing levels and adjusting for market returns often changes one's assessment of financial variables and their impact on stock prices and returns. We are not alone in making this statement, nor is it new. Graham and Dodd (1932); Graham, Dodd, and Cottle (1962); Dimson (1986); Jacobs and Levy (1988 and 2017); Bloch, Guerard, Markowitz, Todd, and Xu (1993); Guerard, Markowitz, and Xu (2015); Haugen (1998); and Levy (1999) list some 25 variables associated with excess returns.

Burton Malkiel, a retired Princeton University Economics Professor, and the author of the outstanding book, *A Random Walk Down Wall Street* (2016), identifies many of the variables used in this chapter as important variables for stock selection. Robert Haugen identified these variables in his work with Baker (1996 and 2010), and his texts on inefficient markets (1998 and 2009) have documented the case for inefficient markets. These variables are referred to as "financial anomalies" and are useful in ranking stocks and building composite models for stock selection and stock return prediction. The stock rankings can be used as inputs to a portfolio construction and management framework. Robust regression is used to create statistically significant stock rankings as well as stock selection models in a very large global stock universe.

Markowitz developed a portfolio construction model to achieve the maximum return for a given level of risk or the minimum risk for a given level of return (1952, 1959, and 1976). You must estimate expected return models, models of covariance, and impose constraints for effective portfolio construction and management. How does you construct and test financial models? We propose and report three testing levels. Level I and Level II testing are discussed in next two sections. Level III testing is discussed in Chapter 6.

4.2 Fundamental Variables for Stock Selection Modeling

We start with a large set of financial variables studied in Bloch, Guerard, Markowitz, Todd and Xu (1993); Ziemba and Schwartz (1993); Lakonishok, Shleifer, and Vishny (1994); Haugen and Baker (1996); and Guerard, Markowitz, and Xu (2015). These variables are used in US, Japanese, non-US, and global universes by these authors.

Let us define the variables tested in this chapter:

EP = earnings per share / price per share

BP = book per share / price per share

CP = cash flow per share / price per share

SP = sales per share / price per share

DP = dividends per share / price per share

REP = current EP / average of 60 previous months EP

RBP = current BP / average of 60 previous months BP

RCP = current CP / average of 60 previous months CP

RSP = current SP / average of 60 previous months SP

RDP = current DP / average 60 previous months DP

FEP1 = one-year-ahead forecast earnings per share / price per share

FEP2 = two-year-ahead forecast earnings per share / price per share

RV1 = one-year-ahead forecast earnings per share monthly revision / price per share

RV2 = two-year-ahead forecast earnings per share monthly revision / price per share

BR1 = one-year-ahead forecast earnings per share monthly breadth

BR2 = two-year-ahead forecast earnings per share monthly breadth

CTEF = equal-weighted FEP1, FEP2, BR1, BR2, REV1, REV2

PM71 = price momentum (price (t-1)/ price (t-7))

PM121 = price momentum (price (t-1)/ price (t-12))

Alpha = regression alpha of monthly security return against market return;

PMTREND = price momentum with market effect removed

STDEV = standard deviation of daily returns

ROE1YR = one-year return on equity

ROE3YR = three-year return on equity

ROE5YR = five-year return on equity

ROA1YR = one-year return on total assets

ROA3YR = three-year return on total asset

ROA5YR = five-year return on total assets

ROIC = return on invested capital

DI = debt issued

DR = debt repurchased

NDR = net debt repurchased

CSR = common stock repurchased

NCSR = net common stock repurchased

ES = corporate exports (Dividends + Interest Paid + Net Equity Repurchased + Net Debt Repurchased)

Many of these variables are included in the Levy (1999) list of financial anomalies. We need to create tests to access the statistical significance of these fundamental, valuation, and momentum variables, see Haugen (1999) and Bloch et al. (1993).

Level I Testing

Traditional Wall Street analysis generally begins with the calculation of the information coefficients (ICs) which measure the Pearson correlation coefficient between the ranked investment strategy, such as the high earnings-to-price, and ranked subsequent total returns on the portfolios of the investment, see Farrell (1997).[1] We refer to the one-month IC statistical significance test as a "level one" or "level I" testing, see Guerard, Markowitz, and Xu (2014) and Guerard, Markowitz, and Xu (2015).

We can carry out IC testing not only for basic variables, but also for expected return models generated by combining several basic variables. For example, the consensus temporary earnings forecast (CTEF) is an equal weighted forecast earnings, forecast earning revisions, and forecast earning breadth. CTEF is a very effective forecast of firm's growth. The US Expected Return (USER) model discussed below is a dynamic combination of ten factors. The weights of factor are determined by an advanced robust regression technique. We apply the same ten factors and the same regression technique to global investable universe to create the Global Expected Return (GLER) model.

While the CTEF, Alpha, and GLER models are not actually used for asset management, these models are extremely highly correlated with our proprietary models. We share weights and regression analyses. We report efficient frontier analyses, which we refer to as Level II analyses in this chapter. Later, in Chapter 6, we report Data Mining Corrections (DMC) testing results, showing that our estimated stock selection models are statistically significant.

We create a set of global portfolios using regression-based expected return over the 2002–2016 time period that offers substantial outperformance of a global stock benchmark by using the Beaton-Tukey bisquare procedure (1974), which is a robust regression technique. Robust regression is useful in analyzing data plagued by the presence of a large number of residuals from ordinary least squares regressions (OLS) that are outside of the 95 percent confidence interval.

The Beaton-Tukey bisquare weighting scheme weighs variables inversely with their OLS residual. The average Beaton-Tukey bisquare weight is normally 0.93 whereas the OLS weight is 1.00. The authors used the Tukey bisquare procedure in SAS in 1993 and continue to use the Tukey bisquare procedure in their current research. In Bloch, Guerard, Markowitz, Todd, and Xu (1993), the Markowitz and the Daiwa Securities Trust Company Global Portfolio Research Department (GPRD) created and tested a large set of 270-plus fundamental variables, including earnings, book value, cash flows, and sales, in Japan and the US over a period of approximately twenty years. The GPRD database was quarterly. In Bloch et al. (1993) univariate variables (single financial ratios) and composite model rankings produced by regression analyses were input to a mean-variance optimization analysis. Portfolios were ranked on the basis of the Sharpe Ratio. Bloch et al. (1993) reported that the fundamental variables generally worked well in both Japanese and US stock markets and the Tukey bisquare regression modeling produced the highest Sharpe Ratios.[2]

Level II Testing

In this chapter, we update the Bloch et al. (1993) variables and include forecasted earnings acceleration variables and price momentum variables modeled in Guerard, Xu and Gultekin (2012); Guerard, Rachev, and Shao (2013); Guerard, Markowitz, and Xu (2014, 2015); and Guerard (2017). The expected returns are inputs to an optimization system that uses traditional mean-variance and mean-variance tracking error at risk portfolio construction techniques as described by equation (7) in Chapter 1. We could use, and have reported, equal active weighting optimization techniques and other optimization processes, see Guerard, Markowitz, and Xu (2015). The regression models used to create expected returns combine well-established fundamental factors, such as earnings, book value, cash flow, and sales, forecasted earnings acceleration, and price momentum factors. These factors are statistically significant in univariate models (tilts) and in multiple-factor models. We briefly review the applied US and Global equity investment research in Guerard, Xu, and Gultekin (2012); Guerard, Rachev, and Shao (2013); and Guerard, Markowitz, and Xu (2014, 2015). Moreover, we test the effectiveness of the earnings forecasting component, CTEF, of the stock selection model. Guerard, Markowitz, and Xu (2014, 2015) reported the effectiveness of the earnings forecasting variable in global markets. We refer to the estimation of efficient frontiers and the statistical significance test of performance of efficient portfolios as Level II testing.

4.3 Fundamental Variables and Regression-Based Expected Returns Modeling

Background

In this section, we discuss financial modeling and its role in stock selection and portfolio construction. Stock selection and security analysis is not a new avenue of research. Graham and Dodd (1934) and Williams (1938) developed the framework for stock analysis well over 75 years ago. In 1990, Harry Markowitz headed the Daiwa Securities Trust Company's Global Portfolio Research Department. The Markowitz team estimated stock selection models, following the low P/E tradition of Graham and Dodd (1934); Graham, Dodd, and Cottle (1962); Graham (1973); and Dreman (1979, 1998), and tested the fundamental valuation variables mentioned in the previous section.[3] The Markowitz team used relative variables, defined as the ratio of the absolute fundamental variable ratios divided by the 60-month averages of the fundamental variable ratios.

The purpose of the Global Portfolio Research Department (GRPD) team was to identify individual financial variables statistically associated with excess returns, use regression analysis to combine these variables into a composite strategy, test the portfolio strategy out-of-sample, and produce portfolios to outperform the market or universe benchmark. Sophisticated regression techniques not only enhanced in-sample F statistics, but also enhanced the Geometric Means of out-of-sample portfolios.

The GPRD test found that:

1. the outlier and multicollinearity-adjusted, regression-based expected returns outperform equally weighted models or ordinary least squares-weighted expected returns
2. the risk model based on historical daily returns is an effective input to the mean-variance optimization analysis
3. optimization with turnover constraint enhanced portfolio returns. That is, limiting quarterly turnover to 10 percent produced an efficient frontier that was higher than the 20 percent turnover curve, which was higher than the 30 percent and 40 percent turnover constraint efficient frontiers
4. survivor-biased free stock universes were important, but not necessary for model development
5. modeling results must be tested using a data mining corrections test (to be discussed in chapter 6)

GPRD was not the only research group addressing Japanese and US stock regression modeling issues. Bill Ziemba and his staff at Yamaichi Research Institute and Guerard with GPRD presented both US and Japanese research at the 1991 Berkeley Program in Finance (BPF) meeting in Santa Barbara. Chan, Hamao, and Lakonishok presented Japanese-only analysis at the Santa Barbara BPF meeting.[4] Ziemba and Schwartz (1993) summarized the Ziemba-Yamaichi major conclusions for Japan's 1979 -1989 period:

- there is a strong low P/E effect once other fundamental variables are eliminated (the pure effect)
- earnings changes and acceleration are very positive and highly statistically significant
- there are strong, individual sales and price to book effects, but the pure effects are significantly reduced
- higher dividend yielding stocks have higher returns
- there is a pronounced small stock effect in Japan, and a small price effect
- sigma and relative strength effects are highly statistically negatively correlated with returns

The strongest Ziemba and Schwartz (1993) effect was the current year earnings estimate minus the previous year's reported earnings, divided by the stock price.[5] In the Ziemba pure effect model of Japan factors, the strongest pure effect variables were the forecasted earnings changes and acceleration, low price-to-earnings, low price-to-book, and the lower log of market capitalization. Guerard presented the initial modeling results of the research project that evolved into the work discussed in Bloch et al. (1993). The Ziemba and Guerard results in Japan and the US were very similar. In the early 1990s, Asian stock research generally meant Japanese stock research because Japanese stocks were a large weight in global and Non-US stock benchmarks.

Chan, Hamao, and Lakonishok (1991) used seemingly unrelated regression (SUR) to model CAPM monthly excess returns of the value-weighted or equal-weighted market index return as functions of EP, BP, CP, and size, where size is measured by the natural logarithm of market capitalization (LS).[6] Betas were simultaneously estimated, and cross-sectional correlations of residuals were addressed. When fractile portfolios were constructed by sorting on the EP ratio, the highest EP quintile portfolio outperformed the lowest EP quintile portfolio, and the EP effect was not statistically significant. The highest BP stocks outperformed the lowest BP stocks. The portfolios composed (sorted) of the highest BP and CP outperformed the portfolios composed of the lowest BP and CP stocks. In the authors' multiple regressions, the size and book to market variables were positive and statistically significant. The EP coefficient was negative and statistically significant at the 10 percent level. Thus, no support was found for the Graham and Dodd low P/E approach.

In the monthly univariate (seemly unrelated regression) SUR analysis, with each month variable being deflated by an annual (June) cross-sectional mean, Chan, Hamao, and Lakonishok (1991) found that the EP coefficient was negative but not statistically significant, the size coefficient was negative but not statistically significant, the book to market coefficient was positive and statistically significant, and the cash flow coefficient was positive and statistically significant. In their multiple regressions, Chan, Hamao, and Lakonishok (1991) report BP and CP variables were positive and statistically significant but EP was not significant.

Applying an adaptation of the Fama-MacBeth (1973) time series of portfolio cross sections to the Japanese market produced negative and statistically significant coefficients on EP and size but positive and statistically significant coefficients for the BP and CP variables. Chan, Hamao, and Lakonishok (1991, p. 1760) summarized their findings: "The performance of the book to market ratio is especially noteworthy; this variable is the most important of the four variables investigated."

Guerard, Takano, and Yamane (1993) published a study that compared mean-variance portfolio construction with a mean-variance of tracking error portfolio construction techniques using the first-section, non-financial stocks of the Tokyo Stock Exchange for the January 1974–December 1990 period. The paper concluded that an asset manager can create portfolios that maximize expected returns and minimize expected portfolio tracking error, rather than just portfolio total risks. Guerard et al. (1993) measured risk using a tracking error concept, the monthly semi-variance versus the benchmark, the TOPIX. Guerard et al. (1993) reported that the Toyo Keizai earnings forecasts enhanced the traditional Markowitz eight-factor model annual portfolio returns by over 250 basis points, a result consistent with Ziemba and Schwartz (1993) findings.[7]

The number of firms covered by analysts substantially increased globally during the 1991–2010 period, particularly in Japan and China. Guerard, Xu, and Gultekin (2012) extended a stock selection model originally developed and estimated in Bloch et al. (1993) by adding price momentum variable, taking the price at time t-1 divided by the price 12 months ago at t-12, denoted PM, and forecasted earnings acceleration variable (CTEF), which is the composite of the consensus (I/B/E/S) analysts' earnings forecasts and analysts' revisions, to the stock selection model. Guerard (2012) used the slightly varied CTEF variable that consists of forecasted earnings yield (EP), earnings revisions (RV1, RV2), and direction of revisions (BR1, BR2), identified as breadth, as created in Guerard, Gultekin, and Stone (1997).[8] Breadth is defined as the number of analysts increasing earnings forecasts, less the number of analysts reducing earnings forecasts, divided by the total number of earnings forecast estimates.

Guerard also reported domestic (US) evidence that the predicted earnings yield is incorporated into the stock price through the earnings yield risk index. Moreover, CTEF dominates the historic low price-to-earnings effect, or high earnings-to-price (EP) effect. Fama and French (1992, 1995, and 2008) presented evidence to support the BP and price momentum variables as anomalies.[9] We test an Alpha strategy in which the market movement is removed from the price momentum variable.[10]

Global modeling for a "global growth specialist," such as McKinley Capital Management, LLC, involves the use of larger weighting of momentum and forecasted earnings acceleration factors. A composite strategy (MQ) is composed primarily of CTEF and PM, as described in Guerard, Chettiappan, and Xu (2010).[11] Guerard, Xu, and Gultekin (2012) referred to the stock selection model as a USER Model. Guerard, Rachev, and Shao (2013) applied the USER Model to a large set of global stocks for the 1997–

2011 time period. Guerard, Rachev, and Shao (2013) referred to the global expected returns model as the GLER Model. USER and GLER are the "Public Forms" of the McKinley Capital Management models.

As with GPRD, McKinley will discuss its public models in great detail, and not its proprietary models. We can estimate an expanded stock selection model to use as an input of expected returns in an optimization analysis.

REG8 Model

We estimate a monthly model, which we call REG8, for the January 2002–December 2015 period.

$$TR_{t+1} = a_0 + a_1 EP_t + a_2 BP_t + a_3 CP_t + a_4 SP_t + a_5 REP_t + a_6 RBP_t + a_7 RCP_t + a_8 RSP_t + e_t \tag{1}$$

where: *EP, BP, CP, SP, REP, RBP, RCP, RSP* are the above defined variables at time t and e is the randomly distributed error term.

REG9 Model

In Guerard and Mark (2003) and Guerard and Schwartz (2007), the authors added CTEF as a ninth factor in the regression, which we do in this chapter. In testing the REG9 model, we use a weighted latent root regression (WLRR) analysis, combining robust regression and multicollinearity corrected techniques on equation (2) to identify variables statistically significant at the 10 percent level. The REG9 model is written as:

$$\begin{aligned} TR_{t+1} &= a_0 + a_1 EP_t + a_2 BP_t + a_3 CP_t + a_4 SP_t + a_5 REP_t + a_6 RBP_t + a_7 RCP_t + a_8 RSP_t \\ &+ a_9 CTEF_t + e_t \end{aligned} \tag{2}$$

REG10 Model

The third regression model is a ten-factor model (REG10) publishes as USER with US stocks and GLER with Global stocks. The GLER model is estimated monthly using a weighted latent root regression (WLRR) analysis, combining robust regression and multicollinearity corrected techniques on equation (3) to identify variables statistically significant at the 10 percent level. REG10 is written as:

$$\begin{aligned} TR_{t+1} &= a_0 + a_1 EP_t + a_2 BP_t + a_3 CP_t + a_4 SP_t + a_5 REP_t + a_6 RBP_t + a_7 RCP_t + a_8 RSP_t \\ &+ a_9 CTEF_t + a_{10} PM_t + e_t \end{aligned} \tag{3}$$

where $\alpha=(a_{1,\dots,}a_{10})$ is the rolling average of normalized coefficients of $\alpha_{t-1},\dots,\alpha_{t-12}$.
The normalization of α_t takes two steps. First, we set non-positive and non-statistically significant components to be zero. Second, the remaining positive significant components are rescaled so that they add up to 1. The 12-month smoothing is consistent with the four-quarter smoothing in Bloch, Guerard, Markowitz, Todd, and Xu (1993).

While EP and BP variables are significant in explaining returns, the majority of the forecast performance is attributable to other model variables, namely the relative earnings-to-price, relative cash-to-price, relative sales-to-price, price momentum, and earnings forecast variables. The weighting results are extremely consistent with McKinley Capital Management being a Global Growth specialist. The CTEF and PM variables accounted 40 percent of the weights in the GLER Model.[12]

We refer to using WLRR on the first eight variables, the Markowitz Model, as REG8. The model produced out-of-sample statistically significant excess returns in the portfolios. We refer to using WLRR on the ten variables, referred to as GLER in our published articles, as REG10. The REG9 and REG10 model produced out-of-sample statistically significant excess returns in the portfolios and the excess returns most often exceeded the excess returns of REG8. The GLER model is very similar to the work by Bob Haugen.[13]

In a recent update of a 2003 study of the CTEF and the REG9 model, Guerard and Mark (2018) report that the CTEF variable continues to be a positive and statistically significant variable at 5 percent level for stock selection in the US. CTEF, REG9, REG10, and USER have statistically significant ICs (Level I tests), active returns, and Level II tests. The active returns and specific returns (stock selection) fall as one expands the model independent variable from CTEF to 8, 9, and 10 variables in the US. Specific returns are not always statistically significant at 5 percent level. Factor Contributions increase because of increased Earnings Yield, Value, and Industry exposures and returns rise, as we will discuss in Chapter 5.

In Non-US portfolios, Guerard and Mark report that the CTEF variable continues to be a positive and statistically significant variable at 5 percent level for stock selection. CTEF, REG9, REG10, and USER, have statistically significant ICs (Level I tests), active returns, Level II tests, and specific returns. Specific returns (stock selection) fall as one expands the models in the Non-US models. Non-US Portfolio Active Returns rise in REG9 and REG10 because of increased Size, Value, and Industry Factor Contributions. REG9, REG10, and GLER are more rewarded by Active Returns in Non-US than in US portfolios. Furthermore, the US portfolios require much higher targeted tracking errors, 7–8 percent to outperform than Non-US portfolios, which require 4-6 percent. These are extremely important differences. Index-huggers in US portfolios running targeted tracking errors of less than 5 percent will produce far lower Specific Returns, most likely negative, than Non-US portfolios. Investors are better served with a passive ETF than they are with an "Active" portfolio that produces negative Specific Returns.

4.4 Why Apply Robust Regression?

The identification of influential data is an important component of regression analysis. The modeler can identify outliers or influential data and re-run the ordinary least squares regressions on the re-weighted data, a process referred to as robust (ROB) regression. In ordinary least squares (OLS), all data is equally weighted. The weights are 1.0. In robust regression one weights the data inversely with its OLS residual; i.e., the larger the residual, the smaller the weight of the observation in the robust regression.

In robust regression, several weights can be used. We will review the Beaton-Tukey (1974) bisquare iteratively weighting scheme. The intuition is that the larger the estimated residual, the smaller the weight. The Beaton-Tukey bisquare, or biweight criteria, for re-weighting observations is:

$$w_i = \begin{cases} (1-(\frac{|e_i|}{\sigma_\varepsilon}/4.685)^2)^2, & if \frac{|e_i|}{\sigma_\varepsilon} \geq 4.685, \\ 0, & if \frac{|e_i|}{\sigma_\varepsilon} < 4.685. \end{cases} \quad (4)$$

The Beaton-Tukey bisquare robust regression weighting scheme is one of eight schemes that were analyzed in Andrews, Bickel, Hampel, Huber, Rogers, and Tukey (1972). Most robust regression modeling techniques produce similar median forecasting errors, which are substantially lower than the OLS median forecasting errors. Many statisticians prefer to describe this formulation as the bisquare, or biweight function, see Maronna, Martin, and Yohai (2006) for an update and enhancement of the models in Andrews, Bickel, Hampel, Huber, Rogers, and Tukey (1972). [14]

Chen (2002) contains a description of PROC ROBUSTREG. The online SAS documentation of the robust regression weighting functions is also excellent. In the next chapter we will show that alternative robust regression weighting functions produce portfolio returns that are not statistically different from one another.

The robust regression takes care of the first major problem: the existence of outliers among the independent and dependent variables. A second major problem is one of multicollinearity, the condition of correlations among the independent variables.[15] The usual approach is using the principal component analysis of the independent variables only. We find that the latent root regression, which uses latent vectors of the correlation matrix of the combined dependent and independent variables to estimate regression parameters, works better in the presence of multicollinearity. The latent roots, $\lambda_0, \lambda_1, \ldots, \lambda_p$, and the latent vectors $\gamma_0, \gamma_1, \ldots, \gamma_p$ of the combined matrix $\begin{pmatrix} Y'Y & Y'X \\ X'Y & X'X \end{pmatrix}$ where $Y = Y_{n\times1}$ is the dependent variable and $X = X_{n \times p}$ are the p independent variables. The latent roots and latent vectors can describe the inverse of the combined matrix

$$\begin{pmatrix} Y'Y & Y'X \\ X'Y & X'X \end{pmatrix}^{-1} = \sum_{j=0}^{p} \lambda_j^{-1} \gamma_j \gamma_j' \qquad (5)$$

For ten independent variables, the dimension X'X is a 10x10 matrix. Multicollinearity is present when one observes one or more small latent roots. If one eliminates latent vectors with small latent roots ($\lambda_j < .30$) and small first component ($\gamma_{0j} < .10$), i.e., the latent vectors with non-predictive multicollinearity,[16] the "principal component" or latent root regression estimator can be written as:

$$\hat{\beta}_{LRR} = \sum_{j=0}^{P} f_j \delta_j \qquad (6)$$

Where

$$f_j = \begin{cases} 0 & \lambda_j < 0.3 and \gamma_{0j} < 0.1 \\ \dfrac{-\eta \gamma_{0j} \lambda_j^{-1}}{\sum_q \gamma_{0q}^2 \lambda_q^{-1}} & otherwise \end{cases} \qquad (7)$$

$\eta^2 = \Sigma(y - \bar{y})^2$, δ_j, γ_{0j} are part of vector γ_j by $\gamma_j = \begin{pmatrix} \gamma_{0j} \\ \delta_j \end{pmatrix}$, and summation index q are over with those latent roots and latent vectors with predictive power. The presence of multicollinearity produces very small latent roots and vectors, referred to as "non-predictive, near-singularities" in the works of Richard Gunst and his research colleagues. The Weighted Latent Root Regression (WLRR) is a two-step procedure. First, use OLS to produce Beaton-Tukey weights of equation (5). Then apply the latent root regression to weighted observations in equations (6) and (7). We use PROC IML in SAS to eliminate the smaller latent vectors and reduce the problem of multicollinearity.

4.5 SAS Robust Regression Estimations

Let us examine several sets of data and analyses. The financial modeling universe is very large, covering over 45,400 firms with income statement and balance sheet data on the FactSet Research Systems, Inc.(FactSet) database. The financial modeling universe also includes about 16,600 stocks covered by financial analysts in the I/B/E/S database, see Guerard, Markowitz, and Xu (2015). Many I/B/E/S stocks have only one analyst and several firms require at least two analysts' coverage (as we do at McKinley Capital).

Many firms set minimum market capitalization levels for investment. If we restrict the investment universe to firms exceeding $100 MM of capitalization and two analysts, the investable universe would be approximately 7500 stocks, globally. An asset manager and clients must balance higher returns with illiquidity to get achievable results.

Information Coefficients

First, we examine the Information Coefficients (ICs) of 7500 global stocks with two I/B/E/S analysts and the largest market capitalization (exceeding 100MM) for the January 2003–May 2015 time period. The ICs are reported in Table 4.1.

Table 4.1: Global Equity ICs

Factor	Annualized Spread Return	IC	IC T-Stat
GLER	22.31	0.056	4.86
CTEF	20.30	0.055	4.71
RV2	12.95	0.041	3.56
PMTREND	14.35	0.039	3.41
BR2	10.63	0.037	3.15
FY1RV3	10.01	0.035	3.03
BR1	10.56	0.035	2.96
FEP1	12.30	0.032	2.73
ALPHA	9.38	0.031	2.69
FEP2	12.21	0.030	2.60
CP	10.50	0.028	2.43
DP	10.55	0.027	2.37
EP	9.70	0.023	2.01
STDEV	-0.81	0.022	1.92
PM71	5.05	0.020	1.71
ROE1YR	1.24	0.017	1.42
ROIC	2.84	0.016	1.39
SP	5.48	0.015	1.32
ROA1YR	-1.07	0.015	1.25
ROA3YR	-0.25	0.014	1.19
ROA5YR	-0.84	0.014	1.16
ROE3YR	0.77	0.013	1.16
NCSR	-0.29	0.013	1.09
ROE5YR	0.85	0.012	1.07
DI	0.30	0.004	0.38
ES	2.82	0.004	0.36
CSR	1.25	0.004	0.34
BP	7.45	0.004	0.31
PM1	-3.62	0.003	0.32
DR	-3.91	0.002	0.14
NDR	-1.29	-0.004	-0.34
CSI	-0.09	-0.011	-0.92

We can see that the ICs support the GLER, CTEF (and its revisions and breadth components), and the Price momentum variables, PMTrend and Alpha. The ICs of the GLER and CTEF variables are highly statistically significant and the annualized spread return, the return differential between the top decile stocks (90–99, the most preferred) less the bottom decile stocks (0–9, the least preferred), exceeds 20 percent and the spreads are statistically different from zero.

4.6 SAS PROC ROBUSTREG with M, S, and MM Estimations

The composite scores are developed by estimating equation 1 (REG8), equation 2 (REG9 or REG9CTEF), and equation 3 (REG10 or GLER). PROC ROBUSTREG offers many options on the robust weighting functions of residuals. The empirical support for weighting is reported in Andrews et al (1972). The Beaton-Tukey Bisquare Function, shown in equation (4) was used in Bloch et al. (1993), and continues to produce highly statistically significant results.

A typical SAS robust regression program looks like:

```
proc robustreg data=dat2 method = <options>;
model TR =var1,..,vark;
run;
```

In REG9, the M_Bisquare procedure, the basis of WLRR, produces statistically significant Active Returns and Specific Returns that are very competitive to the Fair, Huber, and MM_Yohai procedures. The REG10, or GLER, Model has large weights for both price momentum and CTEF models in Bisquare estimates for the Non-US, Global, and EM universes, shown in Tables 4.2a, Table 4.2b, and Table 4.2c, respectively. In general, and in summary, there are not statistically significant differences in REG9 and REG10 and the robust weighting procedures used in the Global, non-US (XUS) or Emerging Markets (EM) universes.

Table 4.2a: Robust Regression of the Global Universe (January 2003–December 2014)

	Portfolio	Active		Specific		Momentum	
	Return	Return	T-Stat	Return	T-Stat	Return	T-Stat
REG10_S_Yohai	13.52%	7.12%	3.28	2.86%	2.50	1.10%	3.61
REG10_LTS	13.47%	7.07%	3.25	2.53%	2.25	0.98%	3.32
REG10_S_Tukey	13.43%	7.03%	3.23	3.00%	2.59	0.95%	3.31
REG10_M_Huber	13.38%	6.97%	3.17	2.49%	2.19	1.08%	3.59
REG10_M_Fair	13.28%	6.88%	3.12	2.79%	2.39	1.04%	3.42
REG10_MM_Yohai	13.25%	6.85%	3.12	2.49%	2.18	0.99%	3.40
REG10_MM_Tukey	13.03%	6.63%	3.00	2.94%	2.55	0.94%	3.30
REG9_S_Tukey	13.02%	6.62%	2.69	3.36%	2.69	-0.32%	-0.76
REG10_M_Bisquare	13.02%	6.62%	3.03	2.35%	2.08	1.02%	3.38
REG9_M_Bisquare	12.78%	6.38%	2.71	3.24%	2.59	-0.33%	-0.86
REG9_MM_Yohai	12.75%	6.35%	2.70	3.08%	2.52	-0.28%	-0.72
REG9_S_Yohai	12.62%	6.22%	2.63	3.12%	2.60	-0.30%	-0.69
REG9_MM_Tukey	12.56%	6.16%	2.56	3.05%	2.46	-0.37%	-0.95
REG9_M_Huber	12.47%	6.07%	2.57	3.19%	2.63	-0.28%	-0.69
REG9_M_Fair	12.38%	5.98%	2.52	3.02%	2.43	-0.32%	-0.81
REG9_LTS	12.28%	5.88%	2.42	2.99%	2.36	-0.30%	-0.75
REG8_M_Bisquare	10.41%	4.01%	1.48	2.38%	1.98	-1.16%	-2.21

Table 4.2b: Robust Regression of the Non-US Universe (January 2003–December 2014)

	Portfolio	Active		Specific		Momentum	
	Return	Return	T-Stat	Return	T-Stat	Return	T-Stat
REG10_M_HUBER	12.53%	6.13%	3.04	3.15%	2.73	1.00%	3.21
REG10_S_YOHAI	12.43%	6.04%	3.00	2.62%	2.20	1.04%	3.20
REG10_MM_TUKEY	12.42%	6.02%	3.02	2.74%	2.33	0.99%	3.18
REG10_M_BISQUARE	12.40%	6.00%	2.94	2.82%	2.38	0.97%	3.13
REG10_M_FAIR	12.38%	5.98%	2.96	2.85%	2.41	1.07%	3.29
REG10_MM_YOHAI	12.34%	5.95%	2.96	2.90%	2.48	0.88%	2.93
REG10_LTS	12.19%	5.79%	2.85	2.71%	2.28	0.93%	3.02
REG10_S_TUKEY	12.12%	5.72%	2.94	2.32%	1.98	1.07%	3.26
REG9_MM_YOHAI	11.98%	5.58%	2.54	3.10%	2.54	-0.25%	-0.67
REG9_MM_TUKEY	11.91%	5.51%	2.55	2.96%	2.54	-0.29%	-0.72
REG9_M_FAIR	11.87%	5.47%	2.52	2.86%	2.47	-0.18%	-0.44

	Portfolio	Active		Specific		Momentum	
REG9_M_BISQUARE	11.76%	5.37%	2.46	2.98%	2.49	-0.34%	-0.88
REG9_S_YOHAI	11.75%	5.35%	2.45	2.99%	2.47	-0.25%	-0.62
REG9_M_HUBER	11.61%	5.22%	2.37	3.06%	2.54	-0.30%	-0.75
REG9_S_TUKEY	11.43%	5.03%	2.38	2.56%	2.20	-0.20%	-0.50
REG9_LTS	11.31%	4.92%	2.25	2.54%	2.14	-0.39%	-1.04
REG8_S_BISQUARE	9.99%	3.60%	1.53	1.62%	1.45	-1.15%	-2.36

Table 4.2c: Robust Regression of the EM Universe (January 2003–December 2014)

	Portfolio	Active		Specific		Momentum	
	Return	Return	T-Stat	Return	T-Stat	Return	T-Stat
REG10_M_BISQUARE	18.96%	9.59%	4.99	6.44%	4.12	0.66%	2.82
REG10_S_YOHAI	18.89%	9.52%	5.18	6.05%	4.22	0.47%	2.28
REG10_MM_TUKEY	18.71%	9.34%	4.86	6.11%	4.09	0.90%	3.25
REG10_M_FAIR	18.69%	9.33%	4.69	6.02%	3.91	1.07%	3.61
REG10_MM_YOHAI	18.59%	9.22%	4.81	6.02%	3.87	0.58%	2.44
REG10_LTS	18.51%	9.15%	4.75	6.10%	3.90	0.37%	1.63
REG10_M_HUBER	18.48%	9.11%	4.79	6.01%	3.91	0.59%	2.51
EM_REG10_S_TUKEY	17.95%	8.58%	4.39	5.30%	3.42	0.76%	3.01
REG9_S_YOHAI	17.76%	8.40%	4.13	5.84%	4.12	-0.97%	-2.64
REG9_M_HUBER	17.29%	7.92%	4.01	5.56%	3.95	-0.89%	-2.58
REG9_M_BISQUARE	17.25%	7.89%	4.12	5.89%	4.31	-0.81%	-2.39
REG9_MM_YOHAI	17.04%	7.68%	3.88	5.35%	3.76	-1.01%	-2.84
REG9_S_TUKEY	17.03%	7.66%	3.90	5.46%	3.90	-0.94%	-2.68
EM_REG9_LTS	16.81%	7.45%	3.73	5.38%	3.74	-1.01%	-2.86
REG9_M_FAIR	16.80%	7.43%	3.81	5.32%	3.89	-0.88%	-2.51
REG9_MM_TUKEY	16.46%	7.10%	3.74	4.98%	3.69	-0.85%	-2.45
REG8_M_BISQUARE	16.32%	6.96%	3.26	5.53%	3.75	-2.10%	-4.59

Let us examine the monthly regressions for Global stocks in May 2015. The Global PROC ROBUSTREG output for REG8 is shown in Output 4.1.

Program 4.1: REG8 Model

```
proc robustreg data=dat2 method = MM (CHIF=tukey);
modelf1mret =RankedEP RankedBP RankedCP RankedSP R_REP R_RBP R_RCP R_RSP;
run;
```

Output 4.1: Robust Regression Analysis of REG8 Model at May 2015

Parameter Estimates							
Parameter	DF	Estimate	Standard Error	95% Confidence Limits		Chi-Square	Pr > ChiSq
Intercept	1	23.4291	1.8006	19.9000	26.9582	169.31	<.0001
RankedBP	1	0.2212	0.0264	0.1694	0.2729	70.05	<.0001
RankedCP	1	0.1315	0.0295	0.0737	0.1892	19.89	<.0001
RankedEP	1	-0.0162	0.0275	-0.0701	0.0377	0.35	0.5559
RankedSP	1	-0.0111	0.0250	-0.0601	0.0378	0.20	0.6559
R_RBP	1	0.1062	0.0307	0.0460	0.1663	11.97	0.0005
R_RCP	1	-0.0671	0.0290	-0.1239	-0.0103	5.36	0.0206
R_REP	1	0.0165	0.0274	-0.0373	0.0703	0.36	0.5468
R_RSP	1	0.1568	0.0312	0.0957	0.2180	25.24	<.0001
Scale	1	31.2624					

Diagnostics Summary		
Observation Type	Proportion	Cutoff
Outlier	0.0000	3.0000
Leverage	0.1576	4.1874

Goodness-of-Fit	
Statistic	Value
R-Square	0.1504
AICR	1677.859
BICR	1736.234
Deviance	1628172

In the Global M estimation with the Bisquare weighting function, REG8, the RSP and RBP, BP, and CP variables are positive and statistically significant at the 10 percent level. The monthly R-Square is 0.1504. The outliers (Leverage Points) are listed in Output 4.1.

The introduction of CTEF in the May 2015 REG9 Tukey Bisquare robust regression, produces a statistically significant coefficient on CTEF, 0.0625, in Output 4.2.

Program 4.2: REG9 Model

```
proc robustreg data=dat2 method = MM (CHIF=tukey);
model f1mret=RankedEP RankedBP RankedCP RankedSP R_REP R_RBP R_RCP R_RSP RCTEF;
run;
```

Output 4.2: Regression Analysis of REG9 Model at May 2015

Parameter Estimates							
Parameter	DF	Estimate	Standard Error	95% Confidence Limits		Chi-Square	Pr > ChiSq
Intercept	1	20.7302	2.0212	16.7687	24.6917	105.19	<.0001
RankedBP	1	0.2149	0.0265	0.1630	0.2667	65.93	<.0001
RankedCP	1	0.1286	0.0294	0.0710	0.1863	19.11	<.0001
RankedEP	1	-0.0313	0.0279	-0.0860	0.0235	1.25	0.2633
RankedSP	1	-0.0196	0.0251	-0.0688	0.0296	0.61	0.4347
R_RBP	1	0.1113	0.0307	0.0511	0.1714	13.14	0.0003
R_RCP	1	-0.0638	0.0289	-0.1205	-0.0071	4.87	0.0274
R_REP	1	0.0211	0.0274	-0.0327	0.0748	0.59	0.4427
R_RSP	1	0.1682	0.0314	0.1067	0.2296	28.76	<.0001
RCTEF	1	0.0625	0.0213	0.0207	0.1042	8.60	0.0034
Scale	1	31.1364					

Diagnostics Summary		
Observation Type	Proportion	Cutoff
Outlier	0.0000	3.0000
Leverage	0.1499	4.3615

Goodness-of-Fit	
Statistic	Value
R-Square	0.1533
AICR	1686.043
BICR	1750.871
Deviance	1621696

The Tukey Bisquare R-square increases to 0.1533.

The Tukey Bisquare REG10 robust regression of May 2015 (see Output 4.3) produces an R-square of 0.2028, with the positive and statistically significant coefficients on BP, CP, RCP, RSP, and CTEF.

Program 4.3: REG10 Model

```
proc robustreg data=dat2 method = MM (CHIF=tukey);
model f1mret=RankedEP RankedBP RankedCP RankedSP R_REP R_RBP R_RCP R_RSP RCTEF
RPM71;
run;
```

Output 4.3: Regression Analysis of REG10 Model at May 2015

Parameter Estimates							
Parameter	DF	Estimate	Standard Error	95% Confidence Limits		Chi-Square	Pr > ChiSq
Intercept	1	46.3098	2.7433	40.9331	51.6866	284.97	<.0001
RankedBP	1	0.1625	0.0258	0.1119	0.2132	39.54	<.0001
RankedCP	1	0.1154	0.0284	0.0597	0.1711	16.49	<.0001
RankedEP	1	-0.0360	0.0270	-0.0888	0.0169	1.78	0.1821
RankedSP	1	-0.0411	0.0242	-0.0886	0.0064	2.87	0.0903
R_RBP	1	0.0326	0.0303	-0.0268	0.0921	1.16	0.2817
R_RCP	1	-0.0869	0.0280	-0.1416	-0.0321	9.66	0.0019
R_REP	1	0.0154	0.0265	-0.0364	0.0673	0.34	0.5596
R_RSP	1	0.1129	0.0305	0.0531	0.1727	13.68	0.0002
RPM71	1	-0.3101	0.0242	-0.3574	-0.2628	164.80	<.0001
RCTEF	1	0.1217	0.0211	0.0803	0.1632	33.13	<.0001
Scale	1	28.9051					

Diagnostics Summary		
Observation Type	Proportion	Cutoff
Outlier	0.0000	3.0000
Leverage	0.1535	4.5258

Goodness-of-Fit	
Statistic	Value
R-Square	0.2028
AICR	1822.172
BICR	1892.621
Deviance	1509499

Why do we need robust regression? Let us examine the OLS estimation of the GLER Model and its diagnostics. The OLS estimation (see Output 4.4) produced an F Statistic of 60.4, which is highly statistically significant. Positive and statistically significant regression coefficients are produced on the BP, CP, RSP, and CTEF variables. The proportion of leverage points reveals outlier issues. Robust regression is appropriate. It is empirically vindicated in Block et al. (1993).

Program 4.4: OLS REG10 Model

```
proc reg data=dat2;
model f1mret=RankedEP RankedBP RankedCP RankedSP R_REP R_RBP R_RCP R_RSP RCTEF
RPM71;
run;
```

Output 4.4: OLS Regression Analysis of REG10 Model at May 2015

Analysis of Variance					
Source	DF	Sum of Squares	Mean Square	F Value	Pr > F
Model	10	404721	40472	60.24	<.0001
Error	2458	1651401	671.84759		
Corrected Total	2468	2056122			

Root MSE	25.92002	R-Square	0.1968
Dependent Mean	50.51964	Adj R-Sq	0.1936
Coeff Var	51.30682		

Parameter Estimates					
Variable	DF	Parameter Estimate	Standard Error	t Value	Pr > |t|
Intercept	1	45.74720	2.61234	17.51	<.0001
RankedBP	1	0.15258	0.02461	6.20	<.0001
RankedCP	1	0.10236	0.02706	3.78	0.0002
RankedEP	1	-0.02668	0.02567	-1.04	0.2986
RankedSP	1	-0.03493	0.02309	-1.51	0.1305
R_RBP	1	0.02366	0.02887	0.82	0.4126
R_RCP	1	-0.07428	0.02662	-2.79	0.0053
R_REP	1	0.01290	0.02520	0.51	0.6087
R_RSP	1	0.10416	0.02907	3.58	0.0003
RPM71	1	-0.27874	0.02300	-12.12	<.0001
RCTEF	1	0.11348	0.02014	5.64	<.0001

What about the Tukey bisquare robust regression REG10 in the EM universe for May 2015? The R-Square statistic of 0.1781 and the leverage proportion factor of 12 percent are extremely similar to the Global REG10 results. The EP, BP, SP, REP, PM71, and CTEF variables are statistically significant in the EM May 2015 robust regression. See Output 4.5.

Program 4.5: REG10 Model (Emerging Market)

```
proc robustreg data=dat2 method = MM (CHIF=tukey);
model f1mret=RankedEP RankedBP RankedCP RankedSP R_REP R_RBP R_RCP R_RSP RCTEF
RPM71;
run;
```

Output 4.5: Regression Analysis of REG10 at May 2015 (Emerging Market)

Parameter Estimates							
Parameter	DF	Estimate	Standard Error	95% Confidence Limits		Chi-Square	Pr > ChiSq
Intercept	1	26.6994	4.5069	17.8660	35.5328	35.10	<.0001
RankedBP	1	0.3016	0.0511	0.2014	0.4018	34.83	<.0001
RankedCP	1	0.0582	0.0503	-0.0405	0.1569	1.34	0.2476
RankedEP	1	-0.1201	0.0499	-0.2178	-0.0223	5.80	0.0161
RankedSP	1	0.0736	0.0448	-0.0141	0.1613	2.71	0.1000
R_RBP	1	-0.1050	0.0553	-0.2133	0.0034	3.60	0.0576
R_RCP	1	-0.0074	0.0486	-0.1026	0.0878	0.02	0.8788
R_REP	1	0.1376	0.0503	0.0390	0.2362	7.47	0.0063
R_RSP	1	0.0721	0.0535	-0.0327	0.1769	1.82	0.1778
RPM71	1	-0.1106	0.0418	-0.1926	-0.0286	6.99	0.0082
RCTEF	1	0.1709	0.0382	0.0961	0.2457	20.03	<.0001
Scale	1	30.1438					

Diagnostics Summary		
Observation Type	Proportion	Cutoff
Outlier	0.0000	3.0000
Leverage	0.1196	4.5258

Goodness-of-Fit	
Statistic	Value
R-Square	0.1781
AICR	599.2354
BICR	658.1780
Deviance	530797.6

4.7 SAS Robust Regression with the Optimal Influence Function

Modern robust statistics minimize a scale measure of residuals insensitive to large residuals, such as the median of the absolute residuals, see Maronna, Douglas, Yohai, and Salibian-Barrera (2019). The least median squares (LMS) estimator was introduced by Hampel (1975) and Rousseeuw (1984). When we use a very large efficiency measure such as 99 percent, large outliers have virtually no influence on the regression estimates. The larger the efficiency, the larger the bias under contamination, and there can be a trade-off between normal efficiency and contamination by outlier bias. The SAS ROBUSTREG procedure in SAS uses an 85 percent efficiency default level as a result of Maronna, Douglas, and Yoha (2006). We use 99 percent because of research conversations with Doug Martin, see Guerard (2017), and the resulting higher portfolio simulation Sharpe Ratios.

Let us update our REG8, REG9, and REG10 analysis with analyses of three MSCI universes: Global, Non-US, and China A-Shares universes for the December 2002–November 2018 time period.

We estimate REG8, REG9, and REG10 for the MSCI All Country World universe, a global universe (GL), as we did earlier in the chapter. We estimate monthly cross-sectionally regression for one-month ahead total returns (F1MRET) for the REG10 components. We note the statistical significance of the BP, SP, CTEF, and PM71 variables in the Global OLS monthly regression estimation for the December 2017 variables, see Output 4.6. The OLS regression is highly statistically significant, having a F Statistic of 9.50. The OLS regression has many observations with studentized residuals exceeding 2.0, denoting potential outliers.

Program 4.6: OLS REG10 Model

```
proc reg data=dat2;
model F1MRET =EP BP CP SP REP RBP RCP RSP CTEF PM71;
   run;
```

Output 4.6: OLS REG10 Global Universe Estimates

The REG Procedure
Model: MODEL1
Dependent Variable: F1MRET
Date=12/29/2017

Number of Observations Read	2499
Number of Observations Used	2494
Number of Observations with Missing Values	5

Analysis of Variance

Source	DF	Sum of Squares	Mean Square	F Value	Pr > F
Model	10	57.79768	5.77977	9.50	<.0001
Error	2483	1510.65669	0.60840		
Corrected Total	2493	1568.45437			

Root MSE	0.78000	**R-Square**	0.0369
Dependent Mean	0.41937	**Adj R-Sq**	0.0330
Coeff Var	185.99347		

Parameter Estimates

Variable	DF	Parameter Estimate	Standard Error	t Value	Pr > |t|
Intercept	1	0.44330	0.01845	24.02	<.0001
EP	1	0.07611	0.13029	0.58	0.5592
BP	1	0.29928	0.05787	5.17	<.0001
CP	1	-0.00936	0.03719	-0.25	0.8012
SP	1	0.27035	0.09907	2.73	0.0064
REP	1	0.21413	0.24916	0.86	0.3902
RBP	1	-0.01933	0.66321	-0.03	0.9768
RCP	1	0.61371	1.05357	0.58	0.5603
RSP	1	-0.01713	0.02182	-0.78	0.4326
CTEF	1	0.04461	0.01656	2.69	0.0071
PM71	1	0.21742	0.07225	3.01	0.0026

```
OLS Global Diagnostics

The REG Procedure
Model: MODEL1
Dependent Variable: F1MRET
```

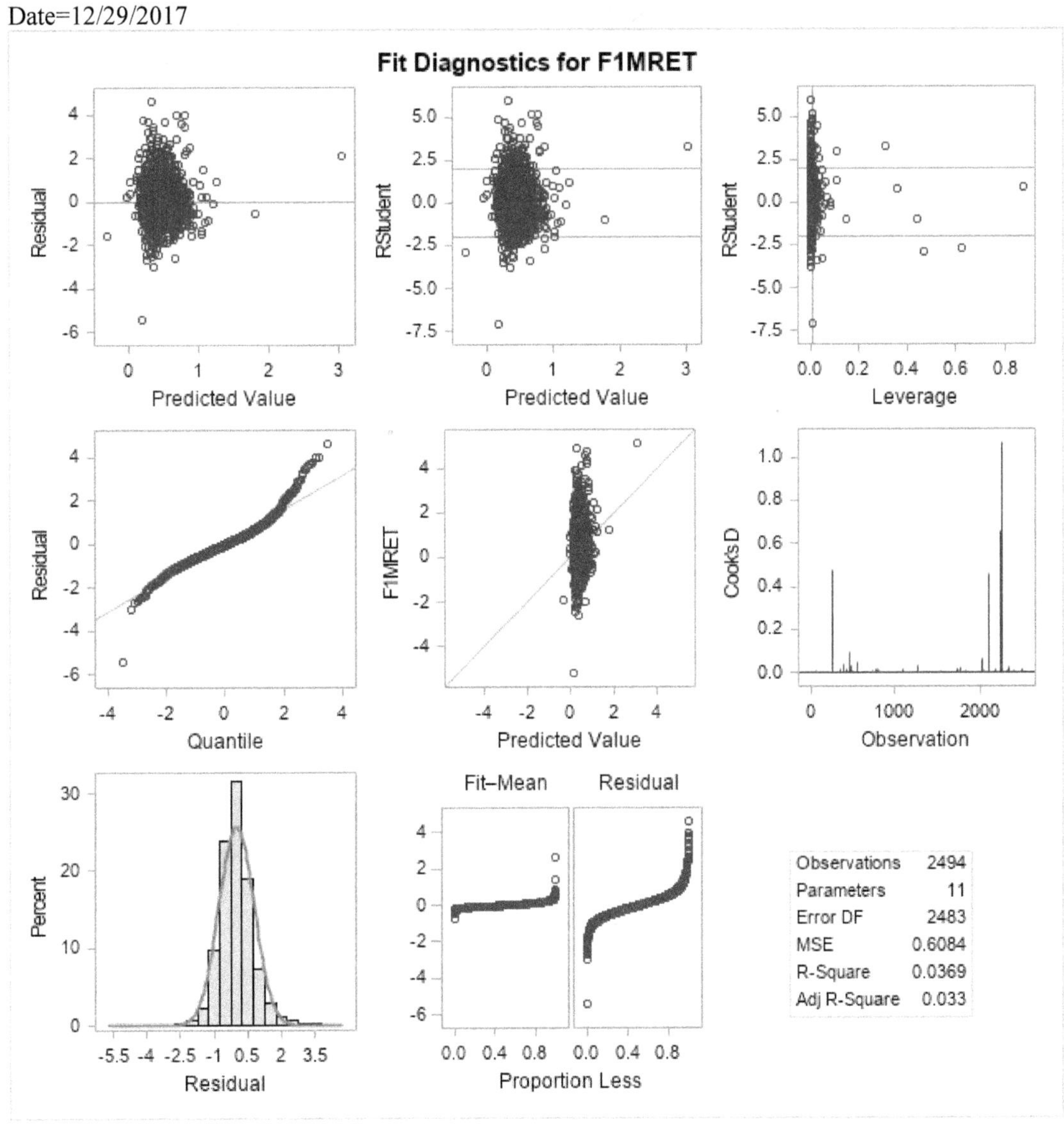

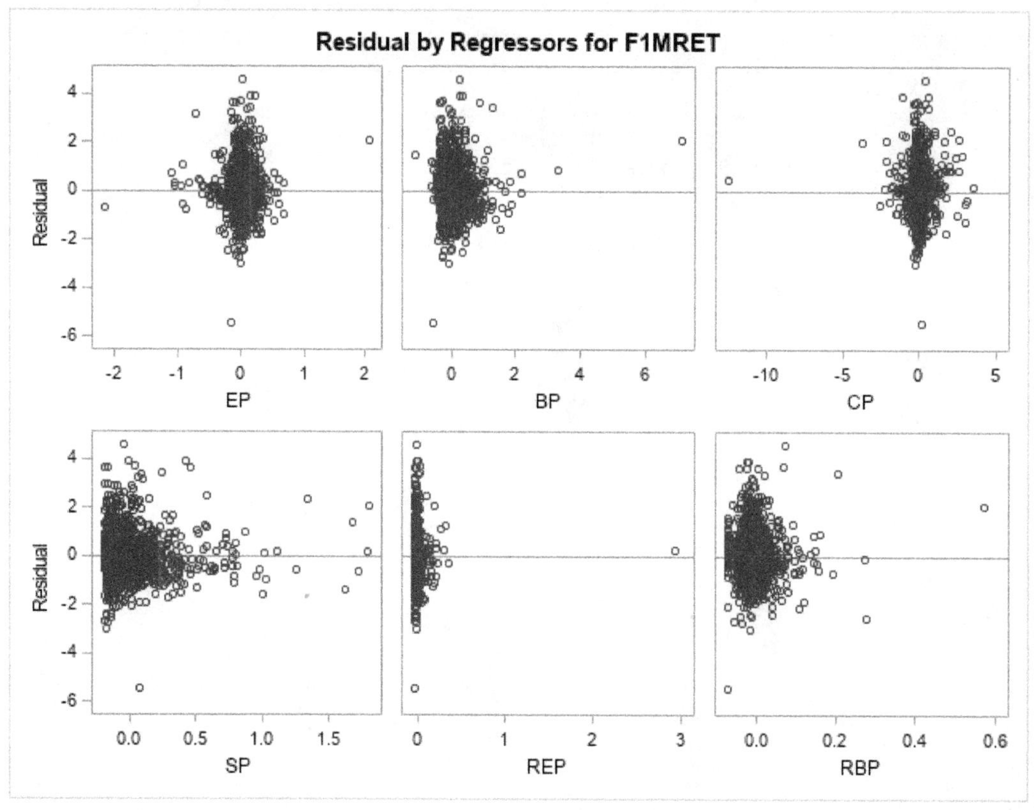

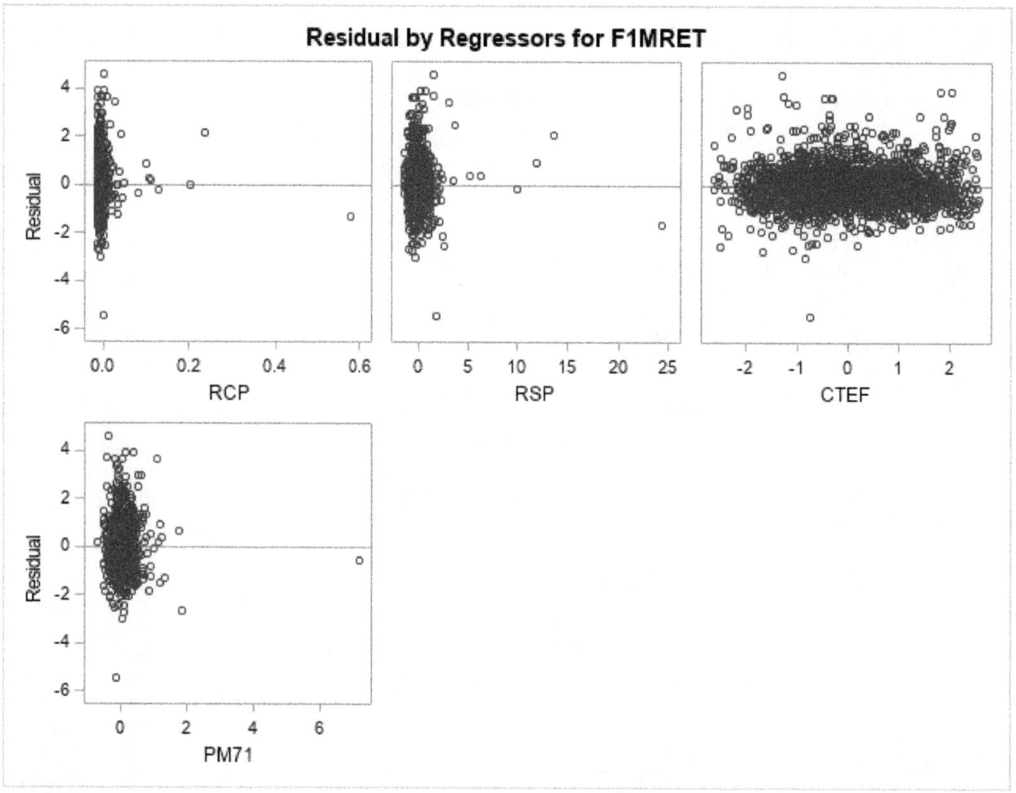

We note the statistical significance of the BP, SP, CTEF, and PM71 variables in the Global Robust monthly regression Tukey OIF99(%) estimation for the December 2017 variables, see Output 4.7. The robust regression estimates identifies many outliers, see Output 4.7.

Program 4.7: REG10 Model (Global)

```
proc robustreg data=dat2 method = MM (CHIF=tukey);
model f1mret=EP BP CP SP REP RBP RCP RSP CTEF PM71;
run;
```

Output 4.7: ROB REG10 Global Tukey OIF99 Estimates

The ROBUSTREG Procedure
Date=12/29/2017

Model Information

Data Set	WORK.NREG
Dependent Variable	F1MRET
Number of Independent Variables	10
Number of Observations	2494
Missing Values	5
Method	MM Estimation

Profile for the Initial LTS Estimate

Total Number of Observations	2494
Number of Squares Minimized	1873
Number of Coefficients	11
Highest Possible Breakdown Value	0.2494

MM Profile

Chi Function	Tukey
K1	7.0410
Efficiency	0.9900

Parameter Estimates

Parameter	DF	Estimate	Standard Error	95% Confidence Limits		Chi-Square	Pr > ChiSq
Intercept	1	0.3970	0.0168	0.3641	0.4299	558.65	<.0001
EP	1	0.0195	0.1181	-0.2119	0.2509	0.03	0.8687
BP	1	0.2367	0.0528	0.1331	0.3402	20.07	<.0001
CP	1	0.0038	0.0338	-0.0624	0.0701	0.01	0.9095
SP	1	0.1671	0.0896	-0.0085	0.3427	3.48	0.0622
REP	1	0.2253	0.2186	-0.2031	0.6538	1.06	0.3026
RBP	1	-1.0863	0.6149	-2.2915	0.1189	3.12	0.0773
RCP	1	0.2078	0.9697	-1.6927	2.1084	0.05	0.8303
RSP	1	-0.0256	0.0207	-0.0662	0.0150	1.53	0.2168
CTEF	1	0.0413	0.0148	0.0122	0.0704	7.75	0.0054
PM71	1	0.1461	0.0657	0.0173	0.2749	4.94	0.0263
Scale	0	0.6765					

Diagnostics

Obs	Standardized Robust Residual	Outlier
28	3.8968	*
35	4.2407	*
58	3.7329	*
66	3.1690	*
216	3.6874	*
287	3.0420	*
288	3.4596	*
291	3.5817	*
311	-3.0421	*
396	-8.1238	*
401	-4.0879	*
424	3.3027	*
425	5.3955	*
459	3.4873	*
472	5.9710	*
492	4.3481	*
504	3.2060	*
515	4.8068	*
528	3.5206	*
563	-3.2189	*
629	3.5578	*
722	3.1139	*
760	4.3578	*
806	3.0998	*
820	3.8916	*
856	-4.4109	*
916	5.1065	*
1088	5.5272	*
1182	3.3546	*
1183	3.2740	*
1271	-3.8114	*
1299	3.0776	*
1322	-3.1298	*
1391	3.5146	*
1463	3.3455	*

Diagnostics

Obs	Standardized Robust Residual	Outlier
1490	3.2988	*
1529	3.7734	*
1571	-3.7506	*
1579	5.3312	*
1648	3.1311	*
1708	-3.4218	*
1724	6.9617	*
1727	3.7307	*
1728	3.1479	*
1759	3.1360	*
1761	3.8040	*
1853	-3.6162	*
1918	3.4588	*
1949	3.8841	*
1966	-3.0137	*
1990	4.4026	*
2001	3.2701	*
2018	5.6349	*
2073	-3.5146	*
2080	4.4113	*
2096	5.3425	*
2117	3.5430	*
2316	4.6978	*
2374	5.6508	*
2397	3.6003	*
2463	3.4206	*
2466	5.9998	*
2487	4.1279	*

Diagnostics Summary

Observation Type	Proportion	Cutoff
Outlier	0.0253	3.0000

Goodness-of-Fit

Statistic	Value
R-Square	0.0202

Goodness-of-Fit	
Statistic	**Value**
AICR	2814.697
BICR	2880.645
Deviance	1279.035

We note the statistical significance of the BP, RCP, CTEF, and PM71 variables in the XUS OLS monthly regression estimation for the December 2017 variables, see Output 4.8. The OLS regression is highly statistically significant, having a F Statistic of 10.46. The OLS regression has many observations with studentized residuals exceeding 2.0, denoting potential outliers.

Program 4.8: OLS REG10 Model (XUS)

```
proc reg data=dat2;
model f1mret=EP BP CP SP REP RBP RCP RSP CTEF PM71;
run;
```

Output 4.8: OLS REG10 XUS Estimates

The REG Procedure
Model: MODEL1
Dependent Variable: F1MRET
Date=12/29/2017

Number of Observations Read	1866
Number of Observations Used	1862
Number of Observations with Missing Values	4

Analysis of Variance

Source	DF	Sum of Squares	Mean Square	F Value	Pr > F
Model	10	99.51406	9.95141	10.46	<.0001
Error	1851	1761.48594	0.95164		
Corrected Total	1861	1861.00000			

Root MSE	0.97552	R-Square	0.0535
Dependent Mean	1.65073E-16	Adj R-Sq	0.0484
Coeff Var	5.909644E17		

Parameter Estimates

Variable	DF	Parameter Estimate	Standard Error	t Value	Pr > |t|
Intercept	1	-0.00046007	0.02261	-0.02	0.9838
EP	1	-0.00625	0.02613	-0.24	0.8111
BP	1	0.14648	0.02923	5.01	<.0001
CP	1	-0.00170	0.02464	-0.07	0.9450
SP	1	0.03855	0.02539	1.52	0.1291
REP	1	0.02001	0.02281	0.88	0.3805
RBP	1	0.03537	0.03172	1.12	0.2649

Parameter Estimates

Variable	DF	Parameter Estimate	Standard Error	t Value	Pr > |t|
RCP	1	0.05488	0.02407	2.28	0.0227
RSP	1	0.02632	0.02903	0.91	0.3647
CTEF	1	0.08021	0.02538	3.16	0.0016
PM71	1	0.07231	0.02487	2.91	0.0037

OLS XUS

The REG Procedure
Model: MODEL1
Dependent Variable: F1MRET
Date=12/29/2017

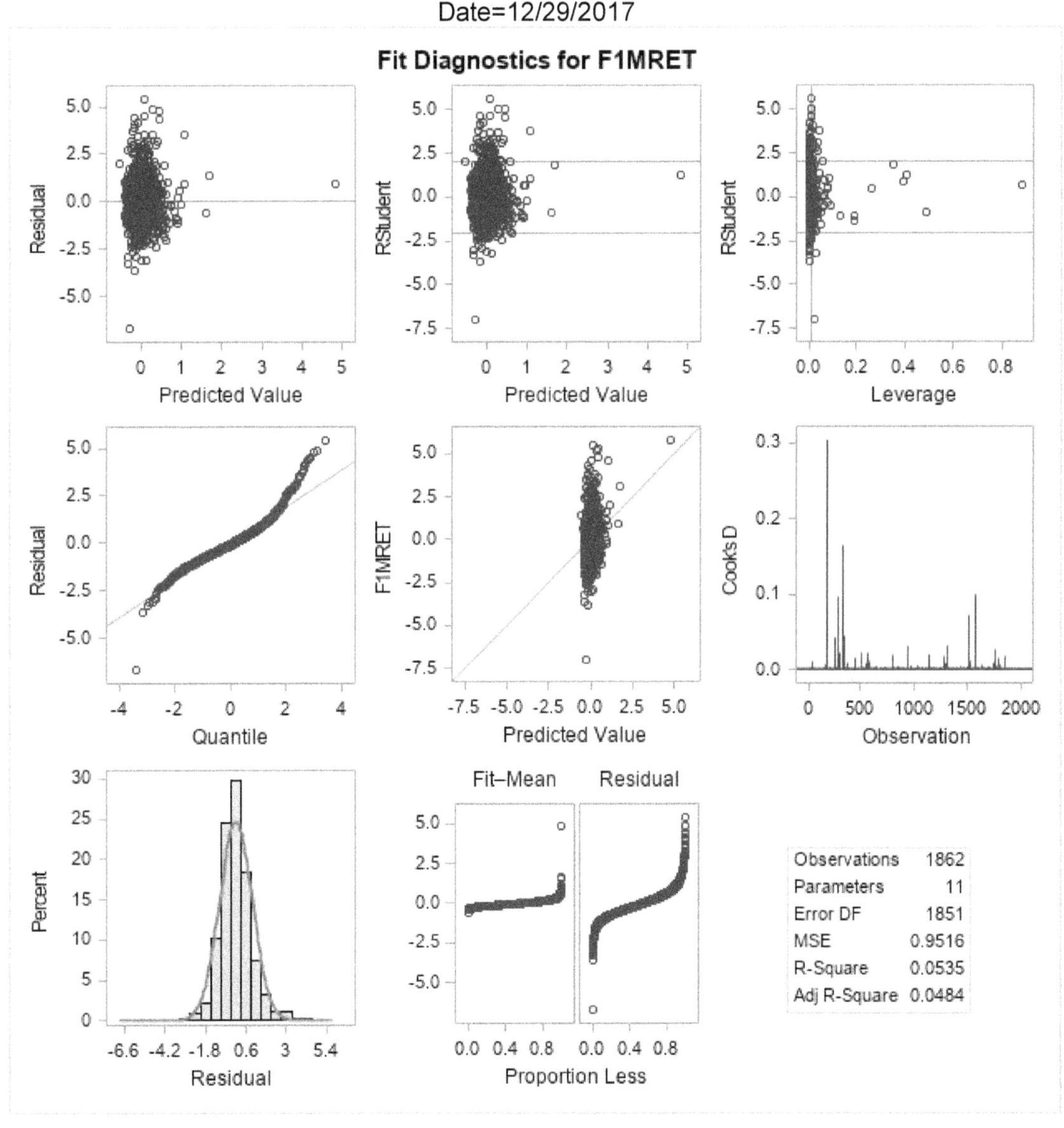

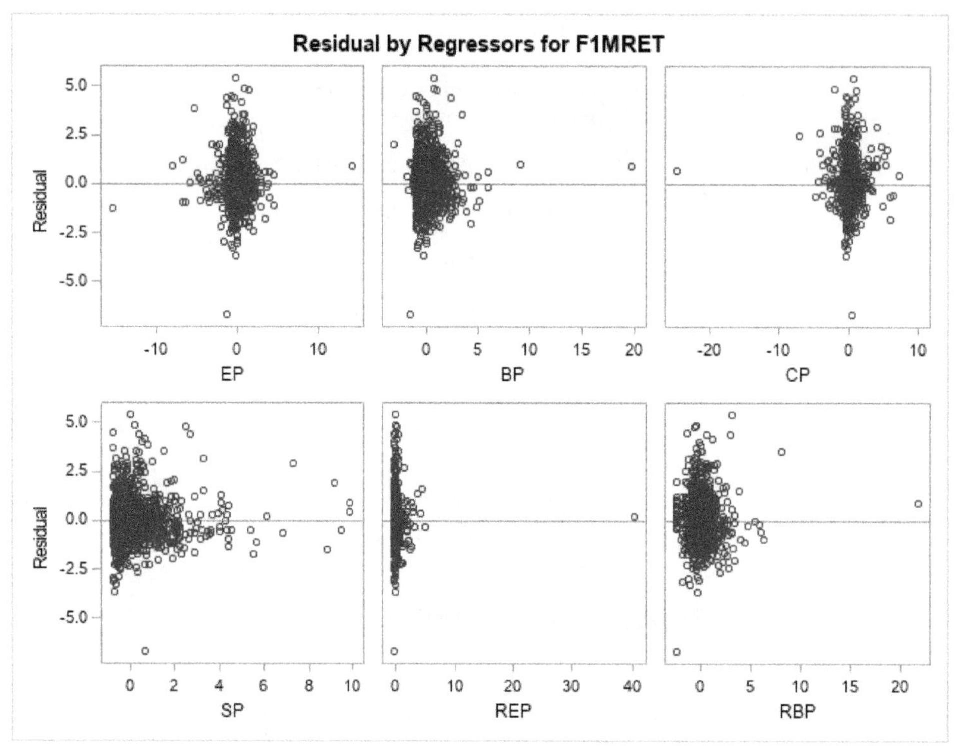

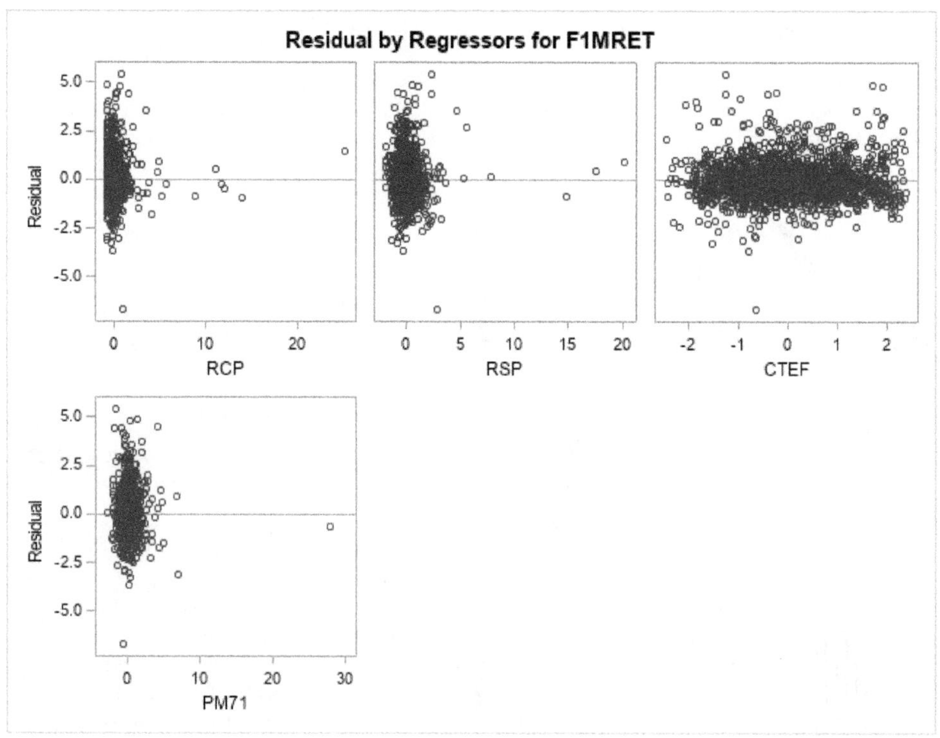

We note the statistical significance of the BP, RCP, RSP, CTEF, and PM71 variables in the XUS Robust monthly regression Tukey OIF99(%) estimation for the December 2017 variables, see Output 4.9. The robust regression estimates identifies many outliers.

Program 4.9: REG10 Model (XUS)

```
proc robustreg data=NREG method = MM (CHIF=tukey);
model f1mret=EP BP CP SP REP RBP RCP RSP CTEF PM121;
run;
```

Output 4.9: ROBUST REG10 XUS Tukey OIF99 Estimates

The ROBUSTREG Procedure
Date=12/29/2017

Model Information

Data Set	WORK.NREG
Dependent Variable	F1MRET
Number of Independent Variables	10
Number of Observations	1862
Missing Values	4
Method	MM Estimation

Number of Observations Read	1866
Number of Observations Used	1862
Missing Values	4

Profile for the Initial LTS Estimate

Total Number of Observations	1862
Number of Squares Minimized	1399
Number of Coefficients	11
Highest Possible Breakdown Value	0.2492

MM Profile

Chi Function	Tukey
K1	7.0410
Efficiency	0.9900

Parameter Estimates

Parameter	DF	Estimate	Standard Error	95% Confidence Limits		Chi-Square	Pr > ChiSq
Intercept	1	-0.1231	0.0251	-0.1724	-0.0738	23.98	<.0001
EP	1	-0.0007	0.0232	-0.0461	0.0447	0.00	0.9760
BP	1	0.1476	0.0259	0.0969	0.1984	32.49	<.0001
CP	1	0.0033	0.0220	-0.0398	0.0465	0.02	0.8794
SP	1	0.0184	0.0226	-0.0260	0.0627	0.66	0.4175
REP	1	0.0242	0.0197	-0.0144	0.0628	1.51	0.2189
RBP	1	0.0111	0.0285	-0.0447	0.0669	0.15	0.6967
RCP	1	0.0558	0.0214	0.0139	0.0978	6.80	0.0091
RSP	1	0.0508	0.0257	0.0004	0.1012	3.90	0.0483
CTEF	1	0.0627	0.0223	0.0190	0.1063	7.90	0.0050
PM121	1	0.0029	0.0005	0.0019	0.0040	31.33	<.0001

Parameter Estimates

Parameter	DF	Estimate	Standard Error	95% Confidence Limits		Chi-Square	Pr > ChiSq
Scale	0	0.8304					

Diagnostics

Obs	Mahalanobis Distance	Robust MCD Distance	Leverage	Standardized Robust Residual	Outlier
1	4.4312	8.8853	*	-0.6396	
2	2.1476	5.9432	*	-0.2449	
3	2.7733	4.9613	*	0.1474	
5	3.6788	7.1012	*	0.5081	
6	3.0910	6.4809	*	-0.2202	
7	3.2118	6.7563	*	-0.2003	
17	2.7715	5.6900	*	1.0389	
18	3.5808	8.0886	*	1.4870	
19	2.8835	7.1671	*	0.0472	
20	4.2941	36.7284	*	-0.3521	
22	2.5014	3.7746		4.0411	*
29	3.5524	7.6497	*	-0.2195	
31	4.0027	8.4349	*	-0.6558	
32	1.7848	5.0130	*	0.2601	
34	2.2808	6.3584	*	-0.5796	
39	4.7465	10.0525	*	1.0192	
41	5.0557	54.5346	*	-0.3846	
42	4.7130	14.8835	*	3.5824	*
45	2.7109	15.3490	*	-0.1335	
49	2.6741	4.4288		3.0049	*
55	2.6130	5.0693	*	0.8469	
56	3.8524	16.1508	*	-0.7594	
57	2.0573	14.8381	*	0.3875	
58	2.8667	5.2350	*	-0.3015	
60	3.1132	7.0999	*	1.2684	
61	3.4518	7.5946	*	-0.3540	
63	2.0750	4.5336	*	0.2206	
64	4.4840	9.7836	*	0.1157	
67	1.4997	5.0494	*	0.5964	
70	6.5625	12.3457	*	0.5502	
75	10.2967	23.9657	*	-0.0832	

Diagnostics

Obs	Mahalanobis Distance	Robust MCD Distance	Leverage	Standardized Robust Residual	Outlier
76	3.1403	10.0074	*	-0.3785	
77	2.2914	4.7939	*	-1.6537	
88	1.7613	4.8870	*	0.0801	
89	1.9840	6.7384	*	0.7154	
94	2.5344	6.8567	*	1.3576	
95	2.0739	5.2424	*	0.3100	
100	1.9926	5.6966	*	-0.1388	
102	2.2790	12.6564	*	-1.4865	
105	2.7365	6.3585	*	0.6870	
112	2.2826	4.9454	*	0.2352	
113	5.4614	23.7822	*	0.5858	
115	2.8385	6.9561	*	-0.1608	
117	2.3332	5.7496	*	-0.0914	
122	3.6766	12.2329	*	0.3747	
136	4.9121	14.3052	*	0.8167	
139	1.7174	5.4701	*	-1.8251	
142	4.0928	9.1107	*	0.7209	
143	2.4153	3.6512		3.5994	*
144	1.8697	4.7775	*	-0.5210	
145	4.0911	17.4127	*	-0.1510	
154	2.7273	4.6477	*	-0.4565	
158	3.0524	4.5405	*	-1.2838	
159	2.2461	4.5297	*	-1.8681	
162	2.0488	6.6436	*	1.1408	
163	2.6663	8.1928	*	0.4020	
164	6.3386	21.6675	*	2.2995	
165	1.8880	5.1402	*	1.3633	
166	1.7866	5.1072	*	1.3644	
172	2.1139	5.0866	*	0.6209	
173	3.9323	11.5408	*	2.2347	
174	5.9812	19.4920	*	1.2191	
176	40.5519	445.7925	*	0.1280	
177	3.5918	7.0583	*	-0.0931	
183	4.0678	7.3819	*	0.2323	
185	2.4698	5.6671	*	2.2610	

Diagnostics

Obs	Mahalanobis Distance	Robust MCD Distance	Leverage	Standardized Robust Residual	Outlier
186	1.9739	5.0896	*	-0.3841	
187	7.1723	33.4182	*	-0.5799	
189	2.2651	7.0007	*	-0.6238	
191	2.6101	4.6586	*	1.6863	
194	1.8332	6.5600	*	0.4551	
195	2.4544	7.4656	*	0.4655	
199	1.5157	5.1725	*	3.0817	*
200	1.9024	5.6712	*	3.4893	*
202	2.6198	8.8543	*	1.8307	
203	1.8774	4.7804	*	3.6823	*
208	2.6833	6.2282	*	-1.7917	
214	2.3430	4.6557	*	0.8071	
215	3.7410	7.6728	*	0.4390	
219	1.6039	2.8965		-3.0537	*
234	1.6782	6.1177	*	0.7145	
235	2.0926	5.9339	*	1.3022	
236	2.2319	4.7364	*	-1.2590	
237	3.6498	8.3354	*	0.4692	
242	5.3301	10.2895	*	2.4176	
247	2.3144	4.6123	*	0.1198	
248	18.8156	51.5402	*	-1.2113	
252	2.5409	5.5136	*	-0.2513	
258	1.9681	4.6521	*	-0.6056	
261	8.3340	15.2647	*	-0.8032	
262	2.2231	6.0902	*	-0.1510	
268	4.5368	7.6853	*	0.7247	
271	1.9408	5.9610	*	1.3216	
274	2.9717	6.7760	*	-0.7974	
275	2.5220	6.7762	*	-0.6416	
283	6.2466	14.2938	*	-8.1605	*
287	2.1582	2.8607		-3.9389	*
290	4.7672	10.0157	*	0.8153	
291	4.8459	10.4676	*	-1.4866	
294	2.3931	4.9002	*	-0.3013	
297	3.5615	5.9385	*	-0.4220	

Diagnostics

Obs	Mahalanobis Distance	Robust MCD Distance	Leverage	Standardized Robust Residual	Outlier
299	2.2017	4.5969	*	3.4721	*
300	2.3412	4.4573		5.4734	*
302	1.1061	5.2815	*	0.7873	
304	4.0295	17.0759	*	-0.3391	
306	4.9346	10.7803	*	0.5957	
307	5.8454	12.7627	*	0.6607	
309	3.6904	10.2726	*	-0.3080	
312	2.3050	4.7745	*	-0.3043	
316	3.2211	6.7834	*	-0.4945	
317	2.2926	4.9041	*	1.7259	
319	1.8231	5.2244	*	-0.8464	
323	3.0072	7.2473	*	-0.1451	
324	4.7398	15.0261	*	1.1480	
325	25.6568	114.0102	*	1.7543	
326	3.0387	8.7098	*	-0.2544	
327	2.3373	4.8730	*	-1.6596	
329	2.2033	5.2839	*	2.7629	
330	2.4903	6.5680	*	0.0591	
331	3.4751	8.2816	*	0.1540	
332	5.3713	18.7742	*	2.0788	
333	3.5571	10.3756	*	0.2428	
334	10.5209	19.2648	*	-2.7979	
335	2.8507	7.1566	*	0.7561	
336	3.0060	4.8888	*	-0.1838	
337	3.6961	12.2736	*	0.0413	
338	3.3471	9.7510	*	5.8326	*
339	6.7282	23.7595	*	-0.6270	
340	3.2082	9.4204	*	0.1763	
346	27.0963	104.2010	*	1.1263	
348	3.5580	6.1169	*	1.7676	
349	4.1084	13.2337	*	1.2800	
351	3.1004	7.2629	*	2.4977	
352	2.5218	7.5650	*	-0.0816	
353	2.4897	5.2142	*	1.5842	
354	4.7516	11.1179	*	-0.3941	

Diagnostics

Obs	Mahalanobis Distance	Robust MCD Distance	Leverage	Standardized Robust Residual	Outlier
355	4.2938	9.7220	*	-0.4760	
358	1.7861	3.7783		4.2964	*
359	1.9425	4.8225	*	-0.4672	
362	1.8830	4.5491	*	-0.1150	
364	1.9593	4.9254	*	2.8639	
365	3.2907	10.6737	*	1.2031	
366	1.8017	5.4349	*	-0.3522	
367	2.6087	5.0540	*	0.9608	
369	5.1777	18.5699	*	3.1728	*
371	2.6246	5.9827	*	-0.9013	
373	2.7845	5.9591	*	-0.4530	
377	2.3849	6.0980	*	1.1116	
378	2.5895	4.6626	*	5.0639	*
380	3.4993	5.6622	*	2.2337	
384	2.7732	5.7096	*	0.3589	
385	2.1060	5.8844	*	3.4955	*
386	1.9686	5.4960	*	0.3432	
387	2.2612	5.1644	*	-1.2227	
388	4.0389	8.4358	*	-0.7855	
389	3.0792	25.5312	*	-1.7683	
390	3.3349	9.0323	*	-0.2804	
392	4.0457	10.1576	*	-0.0052	
397	2.7245	5.2152	*	1.0831	
399	2.2982	5.6128	*	0.6399	
401	5.4899	14.0958	*	0.7406	
403	2.5548	5.4256	*	0.1950	
405	3.2886	6.1929	*	-0.5301	
413	2.7477	5.4095	*	-1.3524	
415	4.9174	13.5517	*	0.2513	
416	7.6799	18.9632	*	-0.5632	
420	2.4535	5.4240	*	-0.9726	
424	4.6952	9.0060	*	-0.0224	
427	2.6054	4.9777	*	-0.8772	
428	4.4140	8.4625	*	0.5830	
434	2.2855	6.4439	*	1.0143	

Diagnostics

Obs	Mahalanobis Distance	Robust MCD Distance	Leverage	Standardized Robust Residual	Outlier
435	2.9573	8.8153	*	-1.0111	
436	2.5782	6.5924	*	0.4389	
437	5.1015	11.9580	*	-0.3825	
440	2.9604	6.8185	*	0.0362	
444	2.4802	5.6398	*	-0.0613	
445	3.0231	5.6761	*	0.2488	
446	2.4980	5.9167	*	-0.3214	
449	4.2436	26.7673	*	-1.4566	
450	15.6421	67.5048	*	-1.1804	
451	2.4428	5.2969	*	-0.9065	
452	2.9460	7.8442	*	-0.8804	
454	3.1779	4.8390	*	2.2647	
455	1.7555	2.3461		3.6608	*
457	2.7167	10.6247	*	0.1089	
462	2.2208	6.3971	*	-0.6141	
464	3.2623	10.6314	*	-0.1344	
465	1.9633	6.0289	*	0.4151	
468	2.6941	6.4086	*	-0.6596	
476	4.9698	14.3371	*	-1.7679	
477	2.5852	6.9962	*	-0.2298	
479	4.0679	7.9097	*	-1.7932	
484	4.6349	12.6303	*	1.1542	
495	2.7640	4.9882	*	0.7695	
499	2.6909	6.7402	*	-0.1437	
501	8.5837	20.5711	*	0.5759	
508	18.8758	48.3701	*	-1.2672	
510	2.8124	6.9635	*	-1.1159	
515	1.0108	1.9025		3.1842	*
516	3.4921	7.1299	*	1.1984	
517	2.6597	5.9357	*	0.5022	
520	2.6178	9.4351	*	-2.5826	
529	3.8903	12.8893	*	-2.3671	
536	2.6262	6.6926	*	-1.4136	
537	2.8822	6.6909	*	-1.0030	
542	4.7942	9.8840	*	-1.4802	

Diagnostics

Obs	Mahalanobis Distance	Robust MCD Distance	Leverage	Standardized Robust Residual	Outlier
549	2.7071	4.3093		4.4492	*
550	2.0476	4.9954	*	-0.8100	
555	2.9813	13.3196	*	0.9969	
558	3.0353	6.8463	*	-0.9225	
559	1.6636	9.9991	*	1.4785	
564	6.7000	14.8997	*	-0.1083	
567	2.2737	5.2264	*	-1.0577	
568	1.7409	6.5118	*	0.0481	
570	10.4590	23.3157	*	2.5604	
572	2.7686	5.5951	*	-2.4250	
573	2.7001	6.0565	*	-1.5151	
575	2.0201	5.0486	*	-1.3862	
577	7.9382	29.2851	*	3.0662	*
586	2.7855	6.6803	*	-0.1891	
587	2.4128	7.6491	*	2.3695	
588	4.3821	10.2649	*	3.9324	*
589	2.1542	5.7768	*	2.0098	
594	2.2458	4.5782	*	0.1892	
605	2.9254	7.2014	*	0.1015	
607	2.2985	6.3821	*	-1.1899	
608	1.2054	2.2172		-4.3564	*
611	2.4284	10.0306	*	-0.0999	
619	2.9870	9.5321	*	0.4727	
624	4.1144	8.8346	*	-0.5147	
625	3.4361	7.5281	*	-0.0347	
627	2.8372	21.2013	*	1.0925	
630	1.3230	7.4049	*	-0.1456	
633	7.0397	12.8750	*	-0.0286	
634	2.6846	5.2388	*	-1.6357	
639	6.0544	10.0282	*	-2.5670	
640	4.7459	9.0758	*	0.4416	
641	4.0417	15.2073	*	0.9288	
642	4.0816	6.7088	*	0.4987	
647	3.5744	8.0837	*	2.4341	
650	2.7902	8.4479	*	1.0140	

Diagnostics

Obs	Mahalanobis Distance	Robust MCD Distance	Leverage	Standardized Robust Residual	Outlier
654	5.9498	27.0923	*	-1.0214	
655	2.0020	3.4412		5.0499	*
656	3.2538	7.1840	*	0.2576	
657	2.1994	4.6476	*	0.4374	
659	3.1772	9.1111	*	-1.7445	
660	6.2979	10.6678	*	0.5739	
665	1.9648	4.7709	*	0.8955	
666	2.9044	6.2179	*	-0.1176	
667	4.0367	8.1798	*	-0.0311	
669	2.8153	5.5044	*	0.6032	
674	2.0833	6.6451	*	-0.3480	
675	1.8951	6.5320	*	-0.1691	
680	2.6058	4.9045	*	-0.9928	
687	2.8984	6.6619	*	-0.1128	
689	1.7625	4.9095	*	-0.0998	
692	5.2950	11.5228	*	1.6149	
693	2.6299	6.3895	*	0.6826	
694	2.6520	4.7714	*	-1.0695	
697	2.9638	5.1266	*	-1.7959	
698	2.9125	7.5936	*	2.4599	
700	2.0674	4.5749	*	0.2908	
703	2.5996	7.5994	*	1.8185	
704	3.9879	10.8978	*	-0.8842	
708	3.2999	10.4611	*	1.1149	
712	3.4948	8.4191	*	-0.5427	
713	3.4091	6.3067	*	-0.6209	
714	2.4233	4.8315	*	0.1989	
719	3.0635	4.7713	*	-1.5053	
723	2.8827	7.0333	*	-1.4880	
724	2.4029	6.9772	*	0.6464	
725	11.9682	37.9731	*	0.7088	
726	4.0363	10.9221	*	-0.1055	
733	2.4730	13.4363	*	1.1963	
738	2.5517	7.1469	*	-0.1774	
742	3.0280	4.9114	*	-2.6369	

Diagnostics

Obs	Mahalanobis Distance	Robust MCD Distance	Leverage	Standardized Robust Residual	Outlier
749	3.1447	6.1649	*	-1.4695	
759	3.2317	6.0752	*	1.7383	
762	2.2862	4.9615	*	-0.2303	
764	2.3139	4.7852	*	-0.8752	
769	5.2323	11.7660	*	-0.9142	
773	2.3784	4.9523	*	0.4839	
777	4.4630	11.5403	*	-0.6765	
778	5.8448	25.8255	*	-0.2747	
779	1.9372	5.6243	*	-0.4859	
781	4.0915	14.0660	*	1.3556	
786	2.9283	5.3912	*	-0.8440	
787	3.7988	17.3396	*	-0.7711	
788	2.1279	4.8514	*	-1.1458	
789	1.6432	5.9574	*	0.5208	
790	1.9309	6.7456	*	-0.2864	
791	2.8397	5.6594	*	-0.2798	
792	2.2880	4.5679	*	-1.0804	
793	3.2439	7.6304	*	1.6415	
794	3.2789	6.6146	*	-0.4893	
795	3.8362	7.5699	*	2.2089	
796	4.3281	12.6895	*	5.3402	*
797	3.1554	8.5066	*	-1.1888	
799	4.2791	10.3103	*	-1.0752	
800	4.6927	11.1871	*	-1.6247	
801	2.1592	7.4039	*	-0.5947	
802	2.9800	5.9004	*	-1.6407	
803	4.2655	10.4675	*	-1.7887	
806	1.2766	6.0626	*	0.7555	
809	2.9809	6.0965	*	-2.6146	
810	3.6934	8.6368	*	-1.8929	
811	1.8356	5.4158	*	-0.1649	
815	1.9688	4.7200	*	-0.0717	
824	2.1807	5.5106	*	1.5470	
825	2.7760	7.8610	*	1.6450	
828	3.2355	6.3105	*	-0.9822	

Diagnostics

Obs	Mahalanobis Distance	Robust MCD Distance	Leverage	Standardized Robust Residual	Outlier
834	3.9172	9.9179	*	0.1700	
835	0.9359	4.8584	*	0.8894	
839	2.5658	7.9847	*	0.2488	
840	2.1370	5.1051	*	0.6477	
843	5.7434	20.5292	*	0.5622	
844	3.0062	6.1087	*	0.3055	
848	3.9297	10.0985	*	1.0176	
850	3.0142	6.0939	*	0.2457	
852	4.2019	40.9998	*	1.6415	
853	5.3645	50.4465	*	1.8802	
854	2.3589	5.4387	*	-0.1945	
856	4.4111	10.1893	*	-0.4632	
857	1.9935	5.8832	*	0.4110	
858	2.5310	7.0921	*	0.3220	
859	2.9123	7.1361	*	0.1570	
861	3.2608	7.1344	*	0.6552	
864	2.0381	4.6725	*	-1.2701	
867	2.4323	5.1112	*	1.0100	
868	1.5455	3.9797		3.3840	*
869	2.1768	4.9631	*	3.3948	*
872	2.5215	5.2707	*	0.3928	
874	3.5055	7.2979	*	-0.4686	
878	4.0250	7.5526	*	-0.6734	
879	5.6413	14.7733	*	-0.7268	
885	2.4561	5.4589	*	-0.4828	
887	3.1417	7.9239	*	-2.2793	
888	3.3180	9.0972	*	-0.0877	
889	4.6822	9.6562	*	0.2645	
891	2.7820	11.2084	*	1.9092	
898	4.5646	10.2000	*	-0.1931	
904	2.2174	5.7559	*	1.0612	
906	3.1900	5.8184	*	-0.5724	
915	2.2669	6.5757	*	-0.7335	
921	2.7752	5.1334	*	-0.2292	
922	1.7361	4.7315	*	1.0546	

Diagnostics

Obs	Mahalanobis Distance	Robust MCD Distance	Leverage	Standardized Robust Residual	Outlier
923	2.5843	6.0806	*	0.3457	
930	3.1251	4.8615	*	-0.1356	
931	2.2312	8.5276	*	1.9458	
934	3.1879	6.5646	*	-0.3634	
937	4.6814	11.5904	*	-0.4273	
938	3.2130	7.7133	*	-0.6822	
941	2.7692	4.6491	*	-0.9164	
944	3.3582	4.8278	*	-3.5432	*
945	3.9993	8.4244	*	0.4554	
950	5.1295	10.8622	*	-0.4926	
951	3.6273	10.2234	*	-0.3279	
952	3.7026	9.4683	*	0.5299	
962	3.9243	10.0108	*	0.2263	
963	1.5947	7.5599	*	0.3820	
967	4.9070	9.5941	*	-2.3226	
968	10.7414	29.0635	*	1.2498	
971	4.3488	13.2046	*	1.1916	
975	3.1067	15.3536	*	-1.3590	
982	2.3282	6.1935	*	-1.5251	
987	2.8154	8.9053	*	-0.2252	
988	3.4770	7.5163	*	-1.2724	
992	1.6623	4.7832	*	0.5779	
994	2.6765	5.2890	*	-0.2065	
1001	3.0040	7.4533	*	-0.2913	
1004	5.5071	11.5005	*	-0.3914	
1005	2.2921	6.1924	*	-0.3368	
1009	4.1339	10.7769	*	0.1647	
1010	2.7870	7.5453	*	-1.5810	
1013	2.4546	9.9637	*	-2.8984	
1014	2.9474	5.8767	*	-1.6377	
1015	2.4983	7.5289	*	0.6943	
1016	3.2701	10.6540	*	-1.2006	
1017	6.2656	11.3521	*	-0.0450	
1022	1.6023	4.9759	*	1.8182	
1023	2.2625	5.5374	*	-1.5089	

Diagnostics

Obs	Mahalanobis Distance	Robust MCD Distance	Leverage	Standardized Robust Residual	Outlier
1027	3.1922	18.1343	*	0.2461	
1030	4.3857	8.4477	*	0.3162	
1033	3.1581	8.1455	*	3.2834	*
1035	2.6848	5.7349	*	1.1777	
1036	3.4555	8.9951	*	0.3669	
1037	5.6270	12.9515	*	1.9119	
1040	2.0278	5.4676	*	0.4227	
1041	4.6040	9.9046	*	0.6777	
1042	4.0293	8.1954	*	-0.5303	
1049	9.3116	16.2154	*	-1.1932	
1051	2.9499	4.2296		-3.0093	*
1052	2.4711	6.1373	*	0.8156	
1053	2.1429	5.2290	*	-0.8821	
1058	2.0602	9.2018	*	0.0591	
1065	4.0934	8.2651	*	-0.5908	
1070	5.1170	11.1214	*	1.4283	
1072	2.7584	5.8655	*	-0.2159	
1073	11.7937	54.2835	*	0.4991	
1075	2.4287	9.6730	*	-0.3983	
1076	4.7884	11.2680	*	0.0459	
1079	5.0676	9.0654	*	-1.2345	
1081	4.8092	10.6769	*	-0.9696	
1082	2.5618	4.4776		3.1825	*
1087	5.2777	11.4127	*	-0.6295	
1094	4.9384	10.3991	*	0.5881	
1103	1.9014	3.8743		3.5376	*
1104	2.4343	5.0762	*	0.9103	
1105	3.1259	7.2114	*	0.6703	
1109	4.0829	10.5238	*	2.1154	
1118	2.5737	5.5895	*	-0.7224	
1119	2.5952	4.8907	*	-0.4398	
1120	2.2381	5.2374	*	-0.3982	
1123	3.2560	9.8145	*	0.2676	
1125	2.7156	18.0332	*	-1.3301	
1127	2.8927	6.3370	*	0.1133	

Diagnostics

Obs	Mahalanobis Distance	Robust MCD Distance	Leverage	Standardized Robust Residual	Outlier
1128	2.3451	4.8895	*	-1.0999	
1130	4.7289	15.8291	*	-0.5071	
1131	3.1127	4.7857	*	0.7098	
1134	2.7698	20.8284	*	-0.2063	
1135	7.2318	20.9435	*	2.9882	
1139	2.4129	5.9323	*	0.1781	
1140	3.4637	12.2115	*	0.5500	
1143	2.7177	6.7724	*	-0.8263	
1153	4.0261	8.2464	*	-0.1387	
1154	2.4289	6.2515	*	-0.4403	
1155	2.8823	7.1909	*	-0.3766	
1157	3.7786	12.6928	*	1.3657	
1158	9.4490	43.4560	*	-0.9191	
1159	5.8942	12.7537	*	0.6286	
1163	2.3255	5.4514	*	0.8303	
1165	2.3765	5.1350	*	-3.4755	*
1175	2.7627	4.9647	*	0.0988	
1178	3.5861	5.6020	*	2.1771	
1183	2.2902	5.1705	*	2.0061	
1189	2.0437	7.2102	*	2.1390	
1191	2.1548	5.1229	*	0.0765	
1197	4.1121	9.2697	*	-0.5960	
1199	2.1519	4.9748	*	0.0800	
1204	4.2527	12.2616	*	0.0901	
1206	3.0035	11.2575	*	-0.4054	
1208	11.9252	51.2810	*	-0.2427	
1209	3.6049	8.7635	*	0.8597	
1213	1.2975	6.8815	*	-1.4516	
1225	4.0435	8.0014	*	-0.2952	
1229	2.6142	8.1978	*	2.3258	
1234	2.7985	5.5378	*	1.1768	
1237	1.5224	6.6645	*	-0.6401	
1239	2.8760	4.9181	*	-0.6615	
1240	10.0269	21.0804	*	-0.3882	
1247	5.8704	15.6499	*	0.5455	

Diagnostics

Obs	Mahalanobis Distance	Robust MCD Distance	Leverage	Standardized Robust Residual	Outlier
1248	3.1584	32.4984	*	0.8290	
1250	3.8855	8.7730	*	-0.4802	
1251	3.3050	7.5174	*	1.3520	
1255	1.9059	14.3879	*	0.9224	
1259	2.4018	5.8237	*	0.9300	
1264	5.5418	10.8338	*	-0.9847	
1267	2.7507	5.6981	*	-2.2914	
1271	2.7280	6.1615	*	-0.8065	
1272	8.1064	20.7646	*	-0.9341	
1273	5.0753	10.3901	*	1.0037	
1275	2.9902	5.1554	*	1.2340	
1279	3.4117	6.3017	*	6.6462	*
1281	2.1919	5.4464	*	1.0280	
1282	2.4535	5.2173	*	3.8847	*
1283	2.4765	5.3408	*	3.2755	*
1288	3.4496	7.9765	*	0.7896	
1289	3.5851	7.8118	*	-1.0606	
1294	2.5153	6.0839	*	1.0146	
1296	12.2123	24.2825	*	1.3093	
1298	2.9102	19.8316	*	-0.0585	
1299	4.2061	8.5975	*	0.3574	
1303	3.6541	7.9060	*	-0.2729	
1304	10.6748	26.9325	*	-0.0602	
1305	1.4395	5.1724	*	3.0086	*
1306	2.6007	5.4266	*	0.9065	
1307	8.0868	17.6376	*	3.8675	*
1308	3.0566	7.3876	*	0.4889	
1310	2.3647	5.6419	*	0.1211	
1311	2.6930	6.3867	*	-1.1394	
1314	1.6900	5.0112	*	0.6584	
1315	4.8136	16.9978	*	0.1679	
1319	5.5621	8.8234	*	0.4466	
1320	2.7747	4.8883	*	-0.2845	
1321	2.3520	4.8449	*	1.1026	
1328	12.8759	58.2938	*	-0.4422	

Diagnostics

Obs	Mahalanobis Distance	Robust MCD Distance	Leverage	Standardized Robust Residual	Outlier
1331	1.9834	5.2710	*	-0.4491	
1339	3.0146	4.7695	*	1.8760	
1340	2.7022	4.9072	*	-1.1786	
1342	2.6213	5.4904	*	0.2120	
1343	2.2912	5.7908	*	0.0409	
1346	3.6663	7.8314	*	0.0296	
1349	5.0229	15.2206	*	1.4193	
1350	2.7493	5.3055	*	-1.0088	
1367	2.2272	4.9474	*	0.3032	
1368	3.8291	15.9693	*	-0.5725	
1369	2.0654	6.2837	*	-0.1687	
1370	1.0216	4.2179		-3.6619	*
1371	2.8473	5.3225	*	-0.8291	
1372	1.4674	6.1850	*	-0.9612	
1373	2.0701	4.7311	*	-0.8512	
1374	1.9002	6.7537	*	0.8753	
1376	2.5548	5.7780	*	-0.0752	
1378	1.5189	6.6952	*	-0.1226	
1379	1.3600	6.3101	*	-0.0338	
1381	2.2558	5.5217	*	0.0465	
1386	1.8036	6.1794	*	-1.0481	
1387	2.2356	4.6479	*	0.5945	
1390	4.5921	11.7209	*	0.2259	
1394	3.0890	29.5370	*	-0.1816	
1396	2.1833	5.8390	*	0.2712	
1397	2.5679	7.0009	*	-0.1208	
1398	2.8643	5.0154	*	1.4098	
1399	3.2562	8.2938	*	-0.8398	
1400	3.9585	7.2065	*	-0.8798	
1401	7.8469	20.1305	*	-1.2158	
1408	2.8758	7.6570	*	0.0423	
1412	3.8369	8.8445	*	0.4476	
1414	3.7631	6.5851	*	1.7452	
1415	2.7180	6.8050	*	0.8848	
1416	3.9440	11.7227	*	-1.2744	

Diagnostics

Obs	Mahalanobis Distance	Robust MCD Distance	Leverage	Standardized Robust Residual	Outlier
1417	2.6116	4.5605	*	-0.0069	
1421	5.2515	11.1972	*	2.5147	
1423	2.7147	5.5001	*	-1.6160	
1424	1.3739	2.6083		3.3286	*
1425	2.9488	6.0673	*	1.6738	
1428	3.1929	5.1776	*	-0.1523	
1432	2.7171	5.7694	*	-1.3231	
1435	7.4632	16.5274	*	-0.1226	
1439	2.5844	4.8589	*	1.9929	
1441	2.1961	4.6237	*	-0.3323	
1443	1.3204	4.8841	*	1.4247	
1460	2.1643	5.5421	*	1.3324	
1463	3.8534	7.1116	*	-0.7597	
1466	4.1110	24.7568	*	-1.0914	
1468	1.9206	5.5045	*	0.3070	
1471	2.1289	4.7027	*	-0.9727	
1473	3.5185	12.2564	*	-1.0463	
1474	1.9775	4.8399	*	-1.1871	
1476	2.4775	9.6897	*	0.6859	
1482	2.2365	4.6358	*	4.2935	*
1483	1.7266	6.1269	*	-0.9524	
1484	6.7527	22.8686	*	-0.6049	
1488	2.9011	6.4250	*	-0.8492	
1490	4.0897	6.9909	*	-2.3237	
1493	4.2571	8.5281	*	2.8624	
1495	3.0070	6.5028	*	-0.1847	
1501	4.2990	9.3686	*	-2.0184	
1503	2.5059	5.7581	*	1.5223	
1508	17.7676	30.9009	*	-0.8139	
1509	9.0332	21.1247	*	4.4111	*
1515	1.9954	5.7478	*	-0.4893	
1519	2.3882	4.8267	*	-0.4798	
1520	3.6635	9.9317	*	1.5609	
1522	2.1692	5.1586	*	0.3132	
1523	3.0697	6.4357	*	-0.2053	

Diagnostics

Obs	Mahalanobis Distance	Robust MCD Distance	Leverage	Standardized Robust Residual	Outlier
1525	5.7514	13.6291	*	0.5081	
1526	3.0139	5.8014	*	-1.7024	
1528	9.5745	21.1817	*	-1.4403	
1530	2.7326	8.1743	*	0.2776	
1531	2.0828	4.5645	*	-0.8090	
1542	2.8861	6.3211	*	0.9597	
1546	4.7863	10.2693	*	-0.7812	
1548	1.9824	7.0663	*	-0.4790	
1552	3.7345	13.8009	*	-0.2749	
1553	1.8709	5.0819	*	1.0987	
1555	1.2759	2.1003		-3.4680	*
1556	2.4597	4.7069	*	0.1746	
1558	2.4699	5.4796	*	0.2622	
1559	3.5098	8.1320	*	-0.3885	
1561	2.4991	5.7009	*	0.6600	
1568	27.8625	64.1924	*	1.4013	
1575	1.9386	4.6238	*	1.6182	
1576	1.7873	5.2978	*	0.0898	
1577	2.3578	5.4927	*	-0.7124	
1579	2.1394	5.3507	*	-0.8437	
1583	3.1439	7.8248	*	-0.5515	
1584	2.4408	7.8598	*	-0.4637	
1587	1.8285	2.4318		3.6253	*
1588	2.2097	6.6208	*	-0.4503	
1591	11.6921	20.5473	*	0.0516	
1595	6.5470	10.8062	*	-0.1682	
1598	5.9234	11.3349	*	-0.9848	
1599	3.5447	7.0470	*	0.8292	
1603	3.5825	8.2810	*	-0.3894	
1605	1.9350	6.0524	*	0.1506	
1607	4.7202	8.4709	*	0.5660	
1609	3.4661	7.2601	*	-0.0953	
1610	3.7152	7.5041	*	-0.0450	
1611	3.2965	6.9609	*	0.1752	
1612	2.0946	5.3508	*	-0.0104	

Diagnostics

Obs	Mahalanobis Distance	Robust MCD Distance	Leverage	Standardized Robust Residual	Outlier
1618	4.2500	10.4408	*	-0.4950	
1622	2.6967	4.7104	*	-0.8675	
1624	22.1491	56.2969	*	0.0442	
1626	4.0570	13.9467	*	0.3451	
1628	2.5422	4.7518	*	-0.1353	
1630	4.9899	10.7950	*	-0.1664	
1635	0.8504	4.5952	*	-0.2287	
1638	3.4040	5.7927	*	-2.1356	
1640	1.0726	5.6905	*	-1.3965	
1643	5.7180	17.9020	*	-1.3491	
1645	1.6025	5.4333	*	-1.3219	
1650	2.4356	6.4554	*	0.7541	
1652	3.0634	24.3877	*	-1.0348	
1654	2.0743	5.2614	*	-0.2571	
1655	2.7102	6.4153	*	0.0531	
1658	1.6858	4.7165	*	0.1398	
1661	3.7846	7.5593	*	1.1572	
1665	2.5479	7.0763	*	0.9742	
1673	2.5537	6.1078	*	-0.0662	
1674	7.9283	20.1381	*	0.2575	
1678	2.5522	5.3235	*	0.7652	
1680	3.9626	34.4053	*	0.6313	
1685	3.6235	7.4243	*	-0.8018	
1691	2.0900	10.0167	*	-0.7623	
1693	8.0277	20.4324	*	-1.0838	
1698	1.6856	5.5791	*	0.2723	
1699	2.4981	5.4398	*	-0.3936	
1702	3.9483	8.1911	*	-0.2915	
1706	2.9439	7.0667	*	-2.4006	
1707	2.7416	5.1734	*	-0.9013	
1708	2.9297	6.3153	*	-0.9649	
1713	2.9691	5.9646	*	-0.8912	
1717	3.2277	6.6489	*	-0.4855	
1721	1.5293	6.1533	*	0.9338	
1723	3.8892	8.0363	*	1.3758	

Diagnostics

Obs	Mahalanobis Distance	Robust MCD Distance	Leverage	Standardized Robust Residual	Outlier
1728	2.0956	10.6323	*	0.5417	
1729	2.5122	4.5824	*	-1.4645	
1730	2.4606	4.9011	*	-0.3313	
1731	1.9438	4.6948	*	-0.5482	
1734	7.1807	21.2782	*	-2.2081	
1735	2.8708	7.8986	*	1.4209	
1736	3.2260	6.9893	*	0.7206	
1737	2.8398	16.1872	*	-2.4916	
1738	2.3207	6.9135	*	0.1196	
1739	2.5723	4.6262	*	0.9105	
1741	2.0426	6.3539	*	0.6030	
1746	5.7946	15.3011	*	4.6403	*
1750	9.0625	23.6374	*	1.6087	
1754	6.7252	15.9938	*	-2.0070	
1755	4.5959	46.4903	*	0.5947	
1756	2.9722	8.7829	*	-0.0808	
1757	8.2759	16.1957	*	0.0379	
1762	2.1433	5.0095	*	-0.3053	
1764	2.5097	20.0946	*	0.4127	
1766	1.9748	4.9518	*	-0.7485	
1769	3.4790	5.7935	*	-2.8784	
1770	2.2320	5.5315	*	0.4322	
1771	4.1569	10.6025	*	-2.7006	
1777	2.1868	5.9620	*	-0.5747	
1778	2.1356	4.7329	*	-0.4840	
1779	2.3122	5.1347	*	1.6458	
1781	3.9248	6.7998	*	5.4411	*
1783	5.5368	10.4988	*	1.9616	
1786	2.0044	5.1645	*	0.4186	
1787	2.1452	5.3331	*	0.4398	
1789	2.2043	5.4884	*	-0.8960	
1794	3.8389	9.0480	*	3.6087	*
1796	3.4495	6.4205	*	2.6769	
1797	2.6947	4.8539	*	1.0374	
1799	2.9370	4.6971	*	0.6381	

Diagnostics

Obs	Mahalanobis Distance	Robust MCD Distance	Leverage	Standardized Robust Residual	Outlier
1804	2.6283	5.4584	*	-0.5464	
1809	7.5138	15.6115	*	0.6664	
1811	3.5828	6.7557	*	0.2558	
1814	2.4699	5.1800	*	0.0345	
1815	6.0437	9.9867	*	-2.2614	
1816	4.3709	7.3978	*	-0.9677	
1818	3.0346	5.2232	*	0.7974	
1819	10.5387	22.5203	*	-0.3217	
1825	2.8647	5.7347	*	-0.0792	
1826	2.3799	7.7445	*	0.4397	
1827	6.2275	13.5244	*	-0.4131	
1830	7.8976	13.3549	*	-0.0137	
1833	2.8838	7.6346	*	0.2663	
1834	3.1835	7.0682	*	-1.3667	
1837	1.5920	3.8973		3.4647	*
1840	3.7684	8.1844	*	5.8676	*
1841	2.7842	6.4781	*	0.1253	
1843	4.0992	6.6212	*	1.3030	
1849	3.4511	8.4254	*	-0.6459	
1853	3.7895	6.9332	*	1.4357	
1859	0.9898	2.0731		4.2248	*
1862	1.9007	5.7708	*	0.4789	

Diagnostics Summary

Observation Type	Proportion	Cutoff
Outlier	0.0269	3.0000
Leverage	0.3550	4.5258

Goodness-of-Fit

Statistic	Value
R-Square	0.0399
AICR	2139.976
BICR	2202.593
Deviance	1461.616

We note the statistical significance of the BP, SP, and CTEF variables in the China A-Shares OLS monthly regression estimation for the December 2017 variables, see Output 4.10. The OLS regression is highly statistically significant, having a F Statistic of 22.77. The China A-Shares regression OLS produces the most highly statistically significant model of three universes. The OLS regression again has many observations with studentized residuals exceeding 2.0, denoting potential outliers.

Program 4.10: REG10 Model (China A Shares)

```
proc reg data=dat2;
model f1mret=EP BP CP SP REP RBP RCP RSP CTEF PM71;
run;
```

Output 4.10: China A OLS REG10 Estimate

The REG Procedure
Model: MODEL1
Dependent Variable: F1MRET
Date=12/29/2017

Number of Observations Read	796
Number of Observations Used	794
Number of Observations with Missing Values	2

Analysis of Variance

Source	DF	Sum of Squares	Mean Square	F Value	Pr > F
Model	10	108.58804	10.85880	22.77	<.0001
Error	783	373.33940	0.47681		
Corrected Total	793	481.92745			

Root MSE	0.69051	**R-Square**	0.2253
Dependent Mean	0.19817	**Adj R-Sq**	0.2154
Coeff Var	348.45086		

Parameter Estimates

Variable	DF	Parameter Estimate	Standard Error	t Value	Pr > |t|
Intercept	1	0.17351	0.02820	6.15	<.0001
EP	1	0.37516	0.04145	9.05	<.0001
BP	1	0.06690	0.03515	1.90	0.0574
CP	1	0.01925	0.02282	0.84	0.3991
SP	1	0.05403	0.03042	1.78	0.0761
REP	1	-0.00032962	0.00251	-0.13	0.8956
RBP	1	0.09222	0.11693	0.79	0.4305
RCP	1	-1.46645	1.95463	-0.75	0.4533
RSP	1	-0.05972	0.03436	-1.74	0.0826
CTEF	1	0.10570	0.02607	4.05	<.0001
PM71	1	0.03791	0.04890	0.78	0.4384

```
ChinaA XUS

The REG Procedure
Model: MODEL1
Dependent Variable: F1MRET
Date=12/29/2017
```

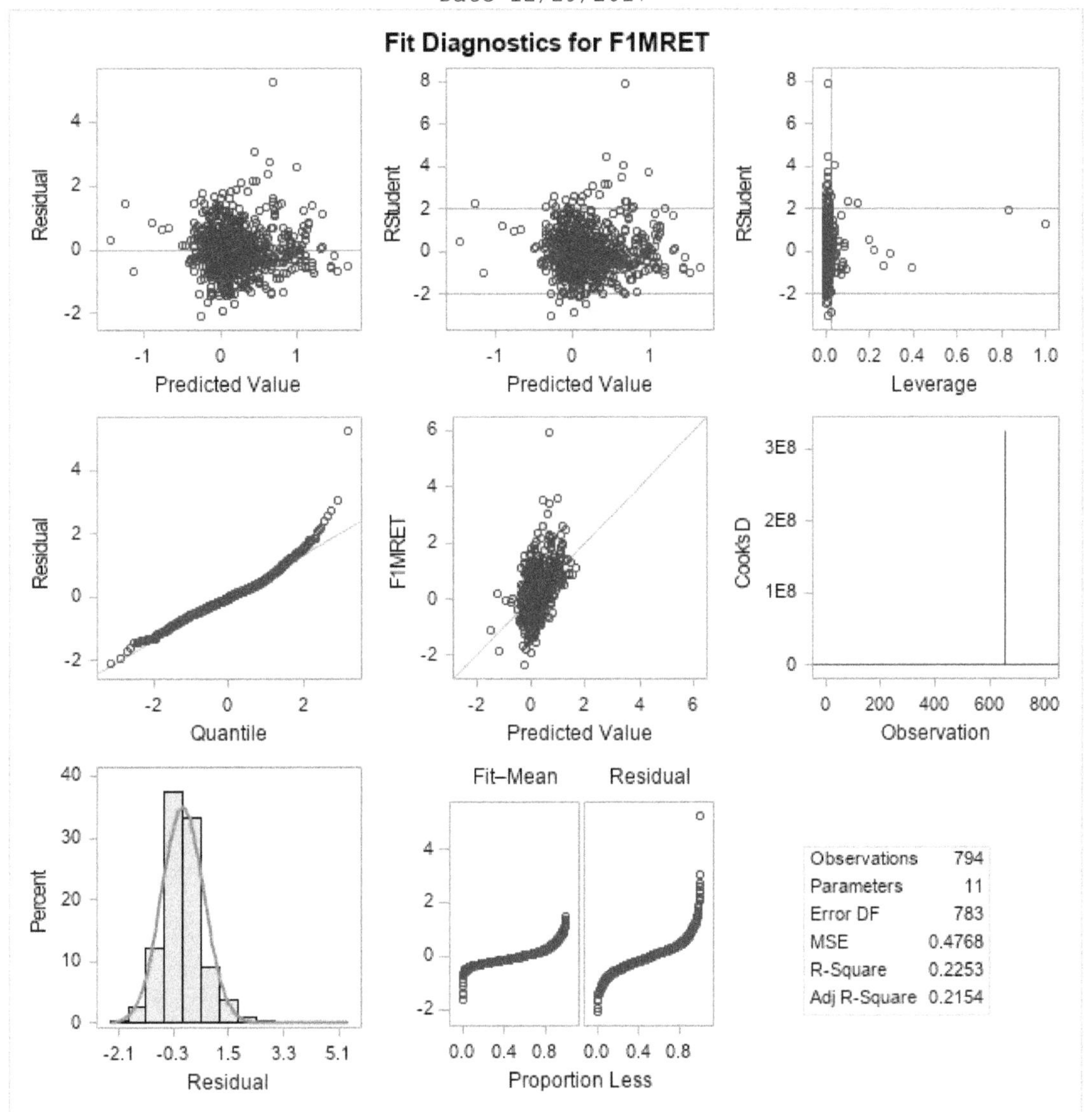

We note the statistical significance of the EP, BP, RSP, and CTEF variables in the XUS Robust monthly regression Tukey OIF99(%) estimation for the December 2017 variables, see Output 4.11. The robust regression estimates identifies outliers, but fewer than the Global and Non-US universes.

Program 4.11: REG10 Model (MSCI ex US)

```
proc robustreg data=nreg method = MM (CHIF=tukey);
model f1mret=EP BP CP SP REP RBP RCP RSP CTEF PM71;
run;
```

Output 4.11: ROBUST ChinaA REG10 Tukey OIF99 Estimates

```
The ROBUSTREG Procedure
Date=12/29/2017

Model Information

Data Set                    WORK.NREG
```

Model Information

Dependent Variable	F1MRET
Number of Independent Variables	10
Number of Observations	794
Missing Values	2
Method	MM Estimation

Profile for the Initial LTS Estimate

Total Number of Observations	794
Number of Squares Minimized	598
Number of Coefficients	11
Highest Possible Breakdown Value	0.2481

MM Profile

Chi Function	Tukey
K1	7.0410
Efficiency	0.9900

Parameter Estimates

Parameter	DF	Estimate	Standard Error	95% Confidence Limits		Chi-Square	Pr > ChiSq
Intercept	1	0.1507	0.0258	0.1002	0.2012	34.22	<.0001
EP	1	0.3482	0.0381	0.2736	0.4228	83.64	<.0001
BP	1	0.0736	0.0319	0.0111	0.1362	5.32	0.0211
CP	1	0.0179	0.0207	-0.0225	0.0584	0.75	0.3850
SP	1	0.0540	0.0275	-0.0000	0.1080	3.84	0.0500
REP	1	-0.0001	0.0022	-0.0045	0.0043	0.00	0.9724
RBP	1	0.0866	0.1063	-0.1217	0.2948	0.66	0.4152
RCP	1	-1.1888	1.7839	-4.6851	2.3076	0.44	0.5052
RSP	1	-0.0771	0.0314	-0.1386	-0.0156	6.04	0.0140
CTEF	1	0.0866	0.0239	0.0397	0.1334	13.12	0.0003
PM71	1	0.0317	0.0454	-0.0572	0.1206	0.49	0.4851
Scale	0	0.6130					

Diagnostics

Obs	Standardized Robust Residual	Outlier
28	-3.4589	*
160	3.4467	*
210	5.0640	*

Diagnostics

Obs	Standardized Robust Residual	Outlier
572	3.0689	*
609	3.6829	*
617	3.0952	*
620	-3.0635	*
737	4.3468	*
738	8.7153	*
761	3.5979	*
773	4.0193	*
775	4.5681	*

Diagnostics Summary

Observation Type	Proportion	Cutoff
Outlier	0.0151	3.0000

Goodness-of-Fit

Statistic	Value
R-Square	0.2001
AICR	863.7335
BICR	917.2064
Deviance	317.0811

Let us conclude with REG8, REG9, and REG10 analysis of the MSCI ex US (the non-US) universe and US Russell 3000 Index constituents for the December 2002–November 2018 time period.

A summary table of non-US results are reported in Table 4.3. We report in Table 4.3 that stock selection is statistically significant, producing 548–796 basis points annually with t-statistics exceeding 2.7, including ITG transactions costs, in the REG8 model estimated for the MSCI ex US universe (XUS) constituents. Total active returns exceed 689–826 in XUS REG8, a result consistent with Guerard and Mark (2018). Total active returns rise with targeted tracking errors (TE), seeing the Information Ratios, IRs, excess returns relative to tracking errors, rise from TE4% to TE8%. No one respects an index-hugger (a TE2% to TE4% manager) unless you are managing $200 billion. The stock-specific effect and total (active) effect are large and statistically significant. As we introduce CTEF in REG9, stock selection diminishes a bit, but is still statistically significant, but total active returns rise. Smaller stocks produce an increase in the size risk premium.[17] As we introduce price momentum in REG10, stock selection diminishes and is no longer statistically significant at the 5 percent level, but total active returns remain large. In summary, in the MSCI ex US universe, EP, CTEF, and REG8, and REG9 produce statistically significant active returns and stock-specific returns.

Table 4.3: Non-US Results

Table 4.3 withTukey OIF99%

Universe: MSCI ex US
Time Period: 12/2002 -11/2018

Portfolios	Sharpe Ratio	Info Ratio	Risk Stock Specific Effect	Risk Stock Specific Effect T-Stat	Risk Factors Effect	Risk Factors Effect T-Stat	Risk Total Effect
REG8_TE4	0.40	0.14	5.48	2.73	1.31	1.72	6.79
REG8_TE6	0.38	0.14	6.94	2.76	0.78	1.71	7.72
REG8_TE8	0.37	0.15	7.96	2.75	0.29	1.73	8.26
REG9_TE4	0.45	0.32	5.38	2.61	2.35	2.80	7.73
REG9_TE6	0.39	0.18	5.49	2.13	2.60	2.93	8.09
REG9_TE8	0.35	0.12	5.52	1.87	2.57	2.80	8.09
REG10_TE4	0.42	0.21	3.07	1.65	3.93	3.62	7.00
REG10_TE6	0.40	0.19	3.03	1.49	5.04	3.66	8.08
REG10_TE8	0.41	0.22	2.95	1.37	5.72	3.47	8.67
EP_TE4	0.53	0.66	7.32	3.70	1.75	2.34	9.07
EP_TE6	0.50	0.50	8.78	3.68	1.03	2.04	9.81
EP_TE8	0.50	0.46	9.78	3.71	1.31	2.22	11.09
CTEF_TE4	0.58	0.79	5.90	3.35	4.02	3.55	9.92
CTEF_TE6	0.61	0.76	6.56	3.33	4.79	3.29	11.34
CTEF_TE8	0.60	0.67	6.82	3.26	5.35	3.07	12.17

Compounded Factor Impact

	Dividend Yield	Earnings Yield	Growth	Leverage	Medium-Term Momentum	Profitability	Size	Value	Volatility	Industry	Country	Currency
REG8_TE4	0.20	0.12	0.12	0.78	-0.91	-0.42	0.95	1.21	-0.70	-0.81	0.67	-0.08
REG8_TE6	0.07	0.07	0.22	0.71	-1.18	-0.70	1.31	1.82	-1.86	-0.56	0.48	0.12
REG8_TE8	0.02	-0.04	0.36	0.62	-0.92	-0.99	1.40	2.25	-2.77	-0.66	0.57	0.30
REG9_TE4	0.19	0.37	-0.01	0.80	-0.20	-0.40	0.84	1.24	-0.49	-0.92	0.76	-0.01
REG9_TE6	0.14	0.36	-0.00	0.77	-0.21	-0.58	1.20	1.89	-1.62	-0.43	0.36	0.59
REG9_TE8	0.12	0.31	0.11	0.72	-0.01	-0.76	1.31	2.39	-2.36	-0.30	0.23	0.77
REG10_TE4	0.15	0.46	-0.04	0.81	0.56	-0.26	0.74	1.12	-0.31	-0.37	0.84	0.18
REG10_TE6	0.10	0.54	-0.01	0.88	0.90	-0.42	1.04	1.65	-1.31	-0.31	0.82	0.94
REG10_TE8	0.19	0.57	0.01	1.02	1.17	-0.41	1.03	1.89	-1.64	-0.48	0.97	1.01
EP_TE4	0.09	0.58	0.06	0.11	0.91	-0.11	0.08	-0.44	-0.22	0.83	1.20	-0.27
EP_TE6	-0.05	0.68	0.07	0.25	0.88	-0.08	0.07	-0.23	-0.37	1.10	1.80	-1.12
EP_TE8	-0.02	0.64	0.07	0.32	0.72	-0.04	0.12	-0.05	-0.49	1.22	2.13	-1.69
CTEF_TE4	0.06	0.76	0.02	-0.22	0.67	-0.05	-0.13	1.78	0.30	0.49	0.28	-0.56
CTEF_TE6	0.05	0.92	0.05	-0.26	0.62	-0.13	-0.03	2.55	0.44	0.66	0.55	-1.60
CTEF_TE8	-0.03	1.00	0.07	-0.23	0.55	-0.16	-0.10	3.07	0.61	0.84	0.54	-2.55

A summary table of US Russell 3000 results are reported in Table 4.4. We report in Table 4.4 that stock selection is statistically significant only in REG8 with a TE of 6 percent producing 166 basis points annually with t-statistics exceeding 1.66, which are statistically significant at the 10 percent level, including ITG transactions costs in the REG8 model estimated for the Russell 3000 universe (R3) constituents. Total active returns are only 98 basis points in the R3 REG8 estimation, a result consistent with Guerard and Mark (2018). The Russell 300 index constituents active returns are far lower than the corresponding non-US portfolios. Total active returns rise with targeted tracking errors (TE), seeing the Information Ratios (IRs) excess returns relative to tracking errors rise from TE4% to TE8% in the REG9 analysis. The stock-specific effect and total (active) effect are larger and more statistically significant than in the R3 REG8 simulations. Thus, as we introduce CTEF in REG9, stock selection increases, is statistically significant, and total active returns rise in REG9. The active returns of the R3 TE8% of 307 basis points is far lower than the corresponding XUS number of 809 basis points, a result consistent with Guerard, Markowitz, Xu, and Wang (2018). Smaller stock produce an increase in the size risk premium. As we introduce price momentum, in REG10, stock selection diminishes and is no longer statistically significant at the 5 percent level, and total active returns diminish. In summary, in the R3 universe, only CTEF and REG9 produce statistically significant active returns and stock-specific returns. Non-US portfolios dominated the R3 portfolio in active and stock-specific returns in the 2003–2018 time period.

Table 4.4: US Russell 3000 Results

Table 4.4 withTukey OIF99%

Universe: Russell 3000
Time Period: 12/2002 -11/2018

Portfolios	Info Ratio	Risk Stock Specific Effect	Risk Stock Specific Effect T-Stat	Risk Factors Effect	Risk Factors Effect T-Stat	Risk Total Effect
REG8_TE4	0.02	-0.06	-0.05	-0.11	0.17	-0.17
REG8_TE6	0.18	1.86	1.66	-0.88	-0.24	0.98
REG8_TE8	0.06	1.77	1.36	-1.53	-0.50	0.24
REG9_TE4	0.12	0.02	0.05	0.31	0.67	0.33
REG9_TE6	0.25	1.35	1.14	0.15	0.55	1.50
REG9_TE8	0.38	3.01	2.04	0.06	0.55	3.07
REG10_TE4	0.12	-0.17	-0.17	0.52	0.89	0.35
REG10_TE6	-0.01	-0.88	-0.62	0.53	0.68	-0.35
REG10_TE8	0.08	-0.42	-0.14	0.85	0.86	0.43
EP_TE4	-0.11	-1.14	-1.39	0.34	0.63	-0.79
EP_TE6	-0.20	-1.56	-1.27	0.11	0.24	-1.45
EP_TE8	-0.41	-3.41	-1.73	-0.46	-0.47	-3.87
CTEF_TE4	0.42	-0.02	0.20	2.03	2.73	2.02
CTEF_TE6	0.42	0.35	0.57	2.59	2.26	2.95
CTEF_TE8	0.52	1.61	1.36	3.30	2.24	4.92

Compounded Factor Impact

	Dividend Yield	Earnings Yield	Growth	Medium-Term Momentum	MidCap	Profitability	Size	Value	Volatility	Industries
REG8_TE4	-0.03	0.10	-0.11	-0.19	-0.21	-0.65	1.93	0.52	-0.80	-0.95
REG8_TE6	0.05	-0.31	-0.15	-0.28	-0.22	-0.74	2.88	0.70	-2.04	-0.89
REG8_TE8	0.10	-0.74	-0.18	-0.41	-0.14	-0.84	3.57	0.90	-3.11	-0.49
REG9_TE4	-0.04	0.53	-0.11	-0.02	-0.17	-0.63	1.82	0.45	-0.61	-1.20
REG9_TE6	-0.01	0.43	-0.13	-0.15	-0.17	-0.69	2.93	0.57	-1.87	-0.93
REG9_TE8	0.08	0.29	-0.15	-0.21	-0.10	-0.62	3.49	0.63	-2.97	-0.38
REG10_TE4	-0.06	0.71	-0.16	0.18	-0.18	-0.53	1.74	0.49	-0.55	-1.30
REG10_TE6	-0.03	0.67	-0.17	0.18	-0.19	-0.50	2.73	0.75	-1.51	-1.42
REG10_TE8	0.11	0.61	-0.17	0.28	-0.13	-0.51	3.33	0.83	-2.38	-0.97
EP_TE4	-0.15	0.68	-0.04	-0.14	-0.21	-0.69	2.10	0.45	-0.75	-1.08
EP_TE6	-0.11	0.55	-0.10	-0.34	-0.16	-0.88	3.61	0.67	-2.15	-1.06
EP_TE8	-0.05	0.28	-0.15	-0.64	-0.08	-0.97	4.62	0.96	-3.18	-0.90
CTEF_TE4	0.00	1.29	0.03	0.72	-0.15	-0.10	1.96	0.04	-0.63	-1.21
CTEF_TE6	0.14	1.66	0.06	1.07	-0.20	0.04	3.20	0.07	-1.63	-1.74
CTEF_TE8	0.26	1.81	0.07	1.41	-0.15	0.30	4.07	0.18	-2.89	-1.51

Price momentum is not useful as an individual variable, see Guerard, Gillam, Markowitz, Xu, Deng, and Wang (2018) but does supply a medium-term momentum risk premium that is highly statistically significant, particularly in non-US, Global, and Emerging Markets during the 2003–2018 time period.

4.8 Summary and Conclusions

In this chapter, we reported evidence confirming the continued relevance of the robust regression models in estimating stock selection models. The original stock selection model of the Markowitz Group of the 1990s is complemented by adding forecasted earnings acceleration and price momentum variables. The Tukey Bisquare procedures and Beaton-Tukey weighted latent root regression technique have continued to produce statistically significant results in the 2003–2015 time period in a global universe. Recent research by Doug Martin with his colleagues produced Maronna, Martin, Yohai, and Salibian-Barrera (2019), which stresses the Tukey and Yohai weightings in the optimal influence function (OIF). We use the Tukey OIF with a 99 percent efficiency parameter and report highly statistically significant excess returns, far in excess of transactions costs for the 2003–2018 time period. We report additional evidence of the continued presence of financial anomalies, consistent with earlier evidence of Guerard and Markowitz (2019). We use robust regression and the most sophisticated of statistical-based risk models and optimization. This chapter guides students and researchers to use SAS statistical techniques that enhance portfolio returns, producing statistically significant Active Returns and Specific Returns.

[1] The strategy in this analysis is ranked, 0–99, where 99 is most preferred.

[2] Since GPRD tested so many models, Markowitz-Xu (1994) developed a data mining corrections test that established that the regression model returns were significantly higher than the average model returns. The data mining test establishes statistically that the excess returns were not due to chance. Data Mining is discussed in the chapter 6.

[3] See also Basu (1977) and Guerard and Stone (1992) for discussions of low price-to-earnings models and other fundamental models such as book value, cash flow, and sales. Bloch, Guerard, Markowitz, Todd, and Xu (1993) reported a set of approximately 270 simulations of GPRD United States and Japanese equity models. The models produced out-of-sample statistically significant excess returns in the portfolios. Haugen and Baker (1996 and 2010) estimated models in the Graham and Dodd framework.

[4] Jacobs and Levy (1988) published several papers in the *Financial Analysts Journal* in the late-1980s. In their U.S. stock study, entitled "Disentangling Equity Return Regularities: New Insights and Investment Opportunities," *Financial Analysts Journal,* May/June (1988), Jacobs and Levy isolated naïve and pure effects of stock selection variables. The authors have great respect for Bruce Jacobs and Ken Levy who endowed a research Center at The Wharton School to study Quantitative Finance. One of the authors of this book, Guerard, spoke at the 2016 annual meeting, the Wharton-Jacobs and Levy Forum, in New York City. William F. Sharpe was named winner of the $80,000 Wharton-Jacobs Levy Prize for Quantitative Financial Innovation for 2015. Mr. Sharpe, co-winner of the 1990 Nobel Memorial Prize in Economic Sciences, was recognized by the Wharton-Jacobs Levy prize committee for his work in returns-based style analysis.
You can read more here: http://www.city-data.com/forum/investing/2578909-bill-sharpe-awarded-wharton-jacobs-levy.html#ixzz56XZY6Ek3.

[5] This is the same form as the FGR variable in Elton, Gruber, and Gultekin (1981).

[6] Chan, Hamao, and Lakonishok (1991) define cash as the sum of earnings and depreciation without explicit correction for other noncash revenue or expenses.

[7] Bill Ziemba served as a referee on Guerard, Takano, and Yamane (1993).

[8] The international I/B/E/S database was created in 1987 and had insufficient data for the Data Mining Corrections test in 1993; hence it was not used in Bloch et al. (1993). Guerard, Takano, and Yamane (1993) used Toyo Keizai earnings forecasts in Japan because of the limitations of the non-U.S. I/B/E/S database. The Toyo Keizai earnings forecasts enhanced portfolio returns by over 200 basis points annually. The GPRD team was well aware of the return-enhancement of I/B/E/S forecasts in U.S. stocks, see Guerard and Stone (1992). The Guerard and Stone research was sponsored by the Institute for Quantitative Research in Finance, the "Q-Group."

[9] The reader is referred to Dimson (1988), Levy (1994), and Ziemba (1993) for a more complete listing of anomalies.

[10] There are many forms of price momentum and the vast majority of the models are very highly correlated, see Brush and Boles (1983) and Brush (2007).

[11] The MQ strategy has worked extremely well since its creation in 2006 in stock selection.

[12] The CTEF and PM variable weights are large relative to the first eight factor eights. The relative variables are "growth" variables such that both the Markowitz model and GLER models plot in the growth boxes of the Zephyr style report. The first four factors are "value" factors.

[13] Haugen (2001) continued the treatment of the Graham and Dodd in his *Modern Investment Theory.* Haugen (1996, 1999) examined 12 of the most important factors in the U.S. equity markets and in Germany, France, Great Britain, and Japan. The book-to-price, earnings-to-price, sales-to-price, and cash flow-to-price variables were among the highest mean payoff variables in the respective countries. Haugen and Baker (2010) published a paper in the Guerard volume to honor Harry Markowitz which updated their models and completely demolished the case for efficient markets. Bob Haugen sadly passed way in early 2013, shortly after his chapter was published.

[14] John Guerard recently presented the MCM robust regression modeling to Doug Martin's (of Maronna, Martin, and Yohai volume fame) Computational Finance and Risk Management Program at the University of Washington. Doug suggested, over a next day (Saturday) research conversation, that Guerard test the Beaton-Tukey bisquare procedure versus the Yohai optimal influence robust procedure at efficiency levels of 85, 95, and 99 percent levels. Doug's suggestions were excellent! The Beaton-Tukey bisquare procedure produces a higher (150 basis point higher but most likely not statistically different) Geometric Mean than the corresponding Yohai procedure. This result is why we run the McKinley Capital "Horse Race" described in Guerard, Gultekin, and Xu (2012). Not all optimization techniques, and risk models, are created equal.

[15] If the independent variables are perfectly correlated in multiple regression, then the (X′X) matrix of equation (1), cannot be inverted and the multiple regression coefficients have multiple solutions. In reality, highly correlated independent variables can produce unstable regression coefficients due to an unstable $(X'X)^{-1}$ matrix. Belsley et al. advocate the calculation of a condition number, which is square root of the ratio of the largest latent root of the correlation matrix relative to the smallest latent root of the correlation matrix. A condition number exceeding 30.0 indicates severe multicollinearity.

[16] Guerard et al (2013) reported the effectiveness of the weighted latent root regression models that made extensive use of correcting outlier-induced collinearities, originally formulated in Webster, Gunst, and Mason (1974).

[17] In Chapter 2, we introduced the reader to the Altman Z bankruptcy prediction score. In the MSCI non-US universe, CTEF outperforms the Altman Z-score for stock selection. The CTEF variable has a 10-year IC of .040 (highly statistically significant, t=6.15) whereas the Altman Z-score has a 10-year IC of 0.02 (statistically significant, t=2.57). In the MSCI EM universe, The CTEF variable has a 10-year IC of .040 (highly statistically significant, t=3.78) whereas the Altman Z-score has a 10-year IC of 0.032 (statistically significant, t=2.13). Both CTEF and the Altman Z-score are statistically significant in stock selection in the McKinley Capital Management investment universes. In the 5-year and 10-year periods, CTEF ranks with the top 3 factors (of over 60 factors) for predictive stock selection, trailing only MQ, the McKinley Capital proprietary model, and the Axioma Medium-Term Momentum for 10-year ICs in the MSCI non-US universe, and only MQ in the EM universe. The Altman Z-score generally ranks in the top third of the factors. CTEF has worked over the majority of the time periods, and been statistically significant, since its publication in 1997.

Chapter 5: The Theory of Risk, Return, and Performance Measurement

5.1 Introduction

The purpose of this chapter is to trace the history of risk and return analysis of stocks and portfolio construction, management, and measurement.[1] Harry Markowitz and William (Bill) Sharpe, have been at the forefront of these topics for the past fifty to sixty years. They have been cited in almost every investment text, many professional journal articles, and their portfolio performance measurements have been cited by the investment practitioner community. We recommend that you refer to the following five reference volumes for this chapter:

- Markowitz, *Portfolio Selection*
- Sharpe, *Portfolio Theory and Capital Markets*
- Rudd and Clasing, *Modern Portfolio Theory: The Principles of Investment Management*
- Elton, Gruber, Brown, and Goetzmann, *Modern Portfolio Theory and Investment Analysis*
- Lee, Finnerty, Lee, Lee, and Wort, *Security Analysis Portfolio Management, and Financial Derivatives*

As we saw in Chapter 1, Harry Markowitz created a portfolio construction theory in which investors should be compensated with higher returns for bearing higher risk. The Markowitz mean-variance analysis (1952, 1956, and 1959) provided the framework for measuring risk, as the portfolio standard deviation.

To many people, both academicians and practitioners, the theory of risk measurement and the risk-return trade-off begins with Markowitz's seminal volume, entitled *Portfolio Selection* (1959). In Chapter 4 of his book, Markowitz introduced standard deviation and variances, and covariance and correlations as measures of dispersion, or risk. Furthermore, Markowitz stated that analysts could and would provide measures of expected returns and variances, but could not reasonably provide estimates of covariance, which could be derived by means of a computer.

Markowitz suggested that the analyst's team could develop graphical relationships between security returns and the change in the index for each of the other stocks in the analysis (pp. 97–98). Thus, we see that Markowitz anticipated the need for quick and efficient calculation of data and the estimation of total and systematic, or market-related, risk. Many readers believe that Markowitz only regarded risk as being measured by variances and covariance, forgetting to read Chapter 9 on the semi-variance, or "down-side risk." Markowitz recognized the existence of skewness (p. 191) and stated quite clearly and correctly that, for a given expected return and variance, investors prefer portfolios with the greatest skewness to the right.

First, this chapter will briefly discuss Markowitz portfolio optimization analysis. Next, the reader is introduced to Capital Market Theory and the relationship between the Capital Market Line and the Security Market Line. Third, we discuss the historical developments and approaches to estimation of Multiple Factor Models (MFMs). We particularly emphasize the BARRA MFM, as it is the risk model most widely used by asset managers. After that, we introduce the reader to alternative risk models, such as Axioma and APT, that can minimize the tracking error of portfolios for asset managers. Finally, we review the Lee, Lee, and Liu (2010) findings on mutual fund performance using the Treynor Index or measures discussed in Treynor-Mazuy (1965).

5.2 Risk and Return and Markowitz Optimization Analysis

Portfolio construction and management, as formulated in Markowitz seeks to identify the efficient frontier, the point at which the portfolio return is maximized for a given level of risk, or equivalently, portfolio risk is minimized for a given level of portfolio return. The portfolio expected return, denoted by $E(R_p)$, is calculated by taking the sum of the security weights multiplied by their respective expected returns:

$$E(R_p) = \sum_{i=1}^{N} w_i E(R_i) \tag{1}$$

The portfolio variance is the sum of the weighted securities covariances:

$$\sigma_p^2 = \sum_{i=1}^{N} \sum_{j=1}^{N} w_i w_j \sigma_{ij} \tag{2}$$

where N is the number of candidate securities, and w_i is the weight for security i such that $\sum_{i=1}^{N} w_i = 1$. This indicates that the portfolio is fully invested, and $E(R_i)$ is the expected return for security i. The Markowitz framework measures risk as the portfolio standard deviation, a measure of dispersion or total risk. One seeks to minimize risk, as measured by the covariance matrix in the Markowitz framework, holding constant expected returns. The decision variables estimated in the Markowitz model are the security weights. The Markowitz model minimized the total risk, or variance, of the portfolio. Investors are compensated for bearing total risk.

The basis form of the Markowitz portfolio optimization analysis can be expressed graphically as shown in Figure 5.1. The decision variables estimated in the Markowitz model are the security weights, subject often to many optimization constraints. One can vary portfolio turnover, the period of volatility calculation, by using five years of monthly data in calculating the covariance matrix, as was done in Bloch et. al. (1993), or one year of daily returns to calculate a covariance matrix, as was done in Guerard, Takano, and Yamane (1993), or 2–5 years of data to calculate factor returns as in the BARRA system, discussed in Menchero et. al. (2010), and later in this chapter.

Figure 5.1: Investment Process

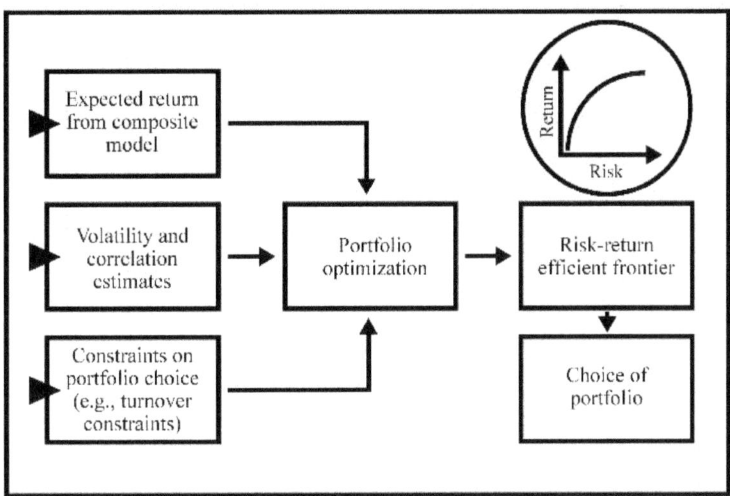

Source: Bloch, Guerard, Markowitz, Todd, and Xu (1993)

Efficient Frontier

The goal in investment research is to operate as closely to the Efficient Frontier was possible. The Efficient Frontier, shown in Figure 5.2, depicts the maximum return for a given level of risk.

Figure 5.2: Iso Risk Contour and Expected Return Line

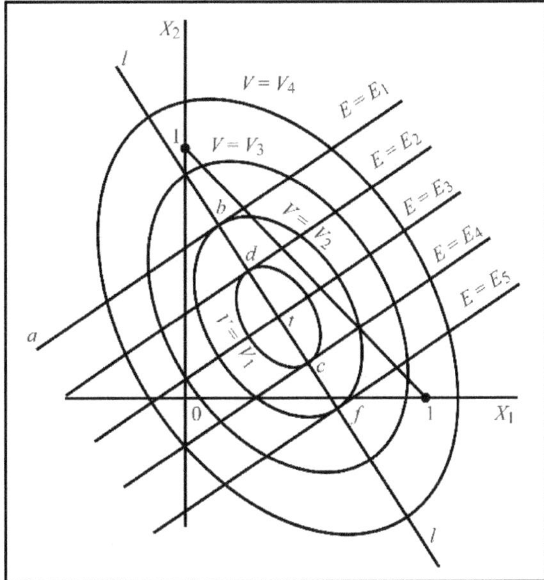

In Bloch et al. (1993), Markowitz and his research department estimated Efficient Frontiers for Japanese stocks for the 1974–1990 period and US stocks for the 1975–1989 period. The Efficient Frontier is shown in Figure 5.3. The upper bound on security weights is a decision variable in which an increase in the upper bound allows the investment manager to exercise more asset selection power and possibly shift out the Efficient Frontier, as in shown in Figure 5.3, in which the upper bound on security weights increases from one percent to two percent.

Figure 5.3: Effect of Holding Upper Bound to Public Model

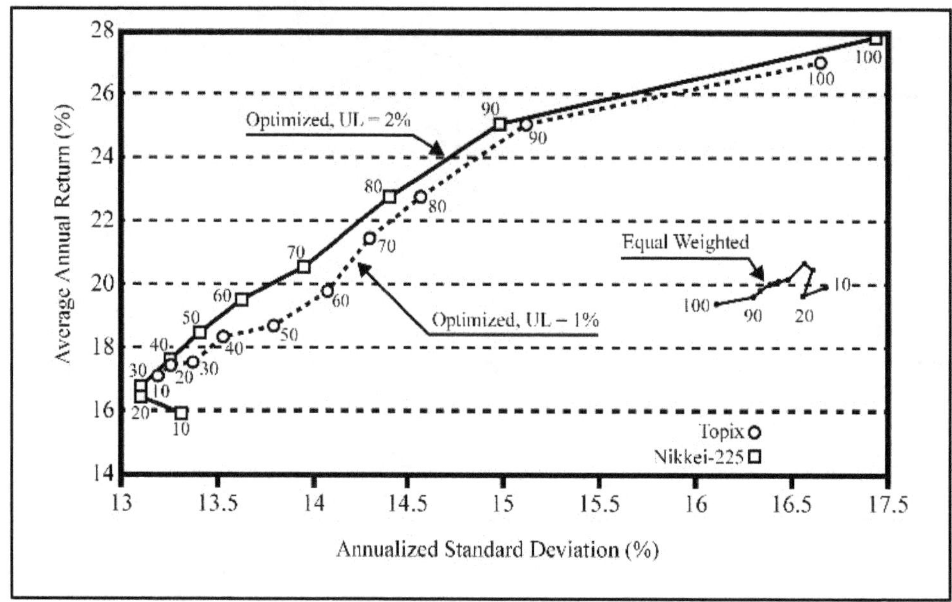

Source: Bloch, Guerard, Markowitz, Todd, and Xu (1993)

The "Public" model trade-off curve, or Efficient Frontier, is shown in Figure 5.3 and the corresponding "Proprietary" model is shown in Figure 5.4. We will more fully develop Figures 5.1 and 5.3 with data in Chapter 6. The trade-off curve contains portfolios of varying risk-return characteristics, with Japanese portfolios producing annualized returns from 17 percent to over 26 percent, annualized.

Figure 5.4: Effect of Holding Upper Bound to Private Model

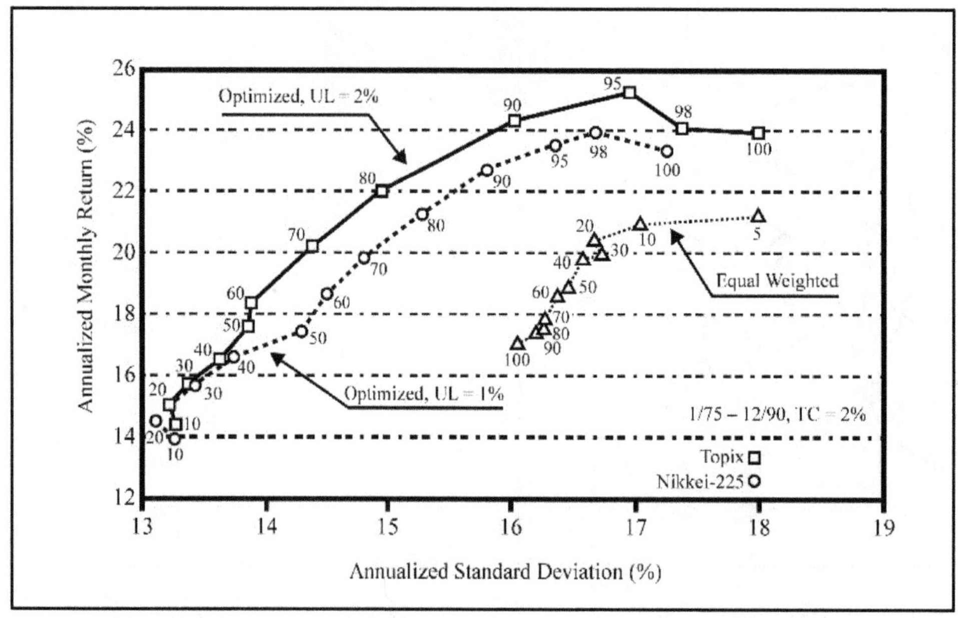

Source: Bloch, Guerard, Markowitz, Todd, and Xu (1993)

An asset manager can create a portfolio with a relatively low return (see the "30" point), or a portfolio with a relatively high return (see the "90" or "95" points). We maximize the Sharpe ratio, the ratio of portfolio excess return (the portfolio return less the index return) relative to the portfolio standard deviation at the "95" point in Figure 5.4. A portfolio that maximizes the Sharpe Ratio is an aggressive portfolio, but it is one where you are compensated most for bearing risk.

A General Form of Portfolio Optimization

The general four-asset linear programming has been written by Martin (1955) as

$$x_1 + x_2 + x_3 + x_4 = 1 \tag{3}$$

$$x_i \geq 0, i = 1, \ldots, 4$$

Expected returns $E = x_1\mu_1 + x_2\mu_2 + x_3\mu_3 + x_4\mu_4$ (4)

Risk $V = x_1^2\sigma_{11} + x_2^2\sigma_{22} + x_3^2\sigma_{33} + x_4^2\sigma_{44} + 2x_1x_2\sigma_{12} + 2x_1x_3\sigma_{13} +$

$$2x_1x_4\sigma_{14} + 2x_2x_3\sigma_{23} + 2x_2x_4\sigma_{24} + 2x_3x_4\sigma_{34} \tag{5}$$

The constrained optimization problem of minimizing risk for given expected return can be reformulated as an unconstrained optimization problem with Lagrange multipliers.

$$\phi = V + \lambda_1\left(\sum_{i=1}^{4} x_i\mu_i - E\right) + \lambda_2\left(\sum_{i=1}^{4} x_i - 1\right) \tag{6}$$

The optimality condition implies

$$\frac{\partial \phi}{\partial x_1} = x_1\sigma_{11} + x_2\sigma_{12} + x_3\sigma_{13} + x_4\sigma_{14} + 0.5\lambda_1\mu_1 + 0.5\lambda_2 = 0$$

$$\vdots$$

$$\frac{\partial \phi}{\partial x_4} = x_1\sigma_{14} + x_2\sigma_{24} + x_3\sigma_{34} + x_4\sigma_{44} + 0.5\lambda_1\mu_4 + 0.5\lambda_2 = 0$$

Markowitz Optimization Equations

The four-asset optimization problem is expressed in a six-equation format by Martin (1955). All partial derivatives are set equal to zero as we did with Markowitz (1959) and the security weights sum to one. The investor is fully invested. The portfolio expected returns is a weighted sum of security expected returns and security weights.

The final set of Markowitz four-asset optimization equations can be written as:

$$\begin{aligned}
&x_1\sigma_{11} + x_2\sigma_{12} + x_3\sigma_{13} + x_4\sigma_{14} + 0.5\lambda_1\mu_1 + 0.5\lambda_2 = 0\\
&x_1\sigma_{12} + x_2\sigma_{22} + x_3\sigma_{23} + x_4\sigma_{24} + 0.5\lambda_1\mu_2 + 0.5\lambda_2 = 0\\
&x_1\sigma_{13} + x_2\sigma_{23} + x_3\sigma_{33} + x_4\sigma_{34} + 0.5\lambda_1\mu_3 + 0.5\lambda_2 = 0\\
&x_1\sigma_{14} + x_2\sigma_{24} + x_3\sigma_{34} + x_4\sigma_{44} + 0.5\lambda_1\mu_4 + 0.5\lambda_2 = 0\\
&x_1\mu_1 + x_2\mu_2 + x_3\mu_3 + x_4\mu_4 = E\\
&x_1 + x_2 + x_3 + x_4 = 1
\end{aligned}$$

Markowitz (1959, pp. 282–83), discusses the Rational Man theory of multi-period investing to maximize terminal wealth. Latane (1959) independently developed his theory of the geometric mean investment criteria at the same time. For n holding period, the annualized holding period return, i.e., the geometric mean (GM), is a more meaningful measurement of return because of the compounding nature. GM is defined as

$$GM = \sqrt[n]{\prod_{t=1}^{n}(1+R_t)} - 1$$

where $\prod$ stands for the product of all 1 plus R_t, return for period t. If we assume return R_t are i.i.d R with expected return E and variance V, then in the long run (n is large)

$$Exp(Ln(1+GM)) = Exp\{\frac{\sum_{t=1}^{n} Ln(1+R_t)}{n}\} \rightarrow Exp(Ln(1+R))$$

$$\approx Ln(1+E) - 0.5\frac{V}{(1+E)^2} \approx E - 0.5V \qquad (7)$$

Where Exp is the expectation and Ln is the natural logarithm.

Equation (7) shows that the portfolio minimizing the variance V for a given level of return will maximize return in the long run. The mean variance approximation of equation (7) worked very well in approximating the GM for the 1-, 4-, 8-, 16-, and 32-stock portfolios in Young and Trent for the 1953–1960 period using 233 stocks. Treynor held that the assumption of risk aversion for a given level of expected returns (performance) in an optimal process led to uncertainty being minimized. Treynor recognized that an optimal portfolio minimized the portfolio variance subject to the constraint that the expected yield was equal to a constant. The use of Lagrange multipliers led to the result that all efficient combinations have the same ratio of risk premium to standard error.[2]

Diagonal Model

The tangency of the locus of efficient combinations with a utility isoquant determined the expected risk premium for the investor in question. In equilibrium, Treynor (2008, p. 56) stated that in his idealized equity market, the risk premium per share of the individual investment is proportional to the covariance of the investment with the total market value of all investments in the market. Thus, in one sentence, Treynor held that the stock risk premium would be proportional to its covariance with the market and made an early statement of the Capital Asset Pricing Model (CAPM). Treynor held the econometric problems of estimating the expected performance and covariance were "outside the scope of this study."[3] Portfolio returns were proportional to the covariances among the stocks, not the total risk of the stocks. The covariances among stocks would be recognized in the coming months as stock betas. Treynor sought to distinguish risk premium for capital budgeting problems between risks that are assumed independent of market fluctuations (the general level of the market), and those that are assumed not to be.

Investments that are risky and independent of market fluctuations are called insurable risks and have a cost of capital equal to the lending rate. Furthermore, the investor holds shares in each equity proportional to the total number of shares available in the market (always positive). In Treynor's 1962 unpublished manuscript, he builds upon Markowitz (1959) and Tobin (1958). The Treynor paper was written contemporaneously with William (Bill) Sharpe (1963 and 1964). Sharpe's 1963 paper, "A Simplified Model for Portfolio Selection" developed the Diagonal Model in which returns of securities are related only through common relationships with an underlying factor; i.e., the stock market as a whole, or Gross National Product, and so on.

The portfolio variance was the slope of the estimated diagonal model squared times the variance of the market. The covariance of two securities was proportional to the product of their respective slopes of the diagonal model. Sharpe noted that the number of estimates for 100 securities fell from 5150 to only 302 using the Diagonal Market. For 200 securities, the number of estimates fell from 2,003,000 to 6002. The time necessary for 100 stock calculations fell from 33 minutes with quadratic programming codes to 30 seconds in the diagonal code. Sharpe concluded that the diagonal model "appears to be an excellent choice for the initial practical applications of the Markowitz technique" (p. 72). [4]

In Sharpe's 1964 paper, "Capital Asset Prices: A Theory of Market Equilibrium under Conditions of Risk," he stated that capital asset prices must change until a set of prices is "attained for which every asset enters at least one combination lying on the capital market line," which is the line between risk and expected return with an intercept of the risk-free rate. Sharpe then demonstrated a consistent relationship between an individual asset's expected returns and systematic risk, the responsiveness of an asset's rate of return to the level of economic activity.[5] Prices adjust until there is a linear relationship between the magnitude of the responsiveness and expected returns (p. 93).

The Capital Asset Pricing Model (CAPM) developed by Treynor (1962, 1999), Sharpe (1964), Lintner (1965), and Mossin (1966) held that investors are compensated for bearing not total risk, but rather market risk, or systematic risk, as measured by a stock's beta. Investors are not compensated for bearing stock-specific risk, which can be diversified away in a portfolio context. A stock's beta is the slope of the stock's return regressed against the market's return. Modern capital theory has evolved from one beta representing market risk to multiple factor risk models (MFMs), many with four or more betas. Investment managers seeking the maximum return for a given level of risk create portfolios using many sets of models that are based both on historical and expectation data. In this chapter, we briefly trace the evolution of the estimated models of risk and show how risk models enhance portfolio construction, management, and evaluation.

5.3 Capital Market Equilibrium

Let us review portfolio theory and capital market equilibrium. Sharpe (1970) discusses capital market theory in which investors purchase or sell stocks based on beliefs or predictions of expected returns, standard deviations of returns, and correlation coefficients of returns. Indeed, all investors have identical expectations of the predictions, known as homogeneous beliefs. Investors seek to maximize return while minimizing risk. Investors can lend and borrow as much as they desire at the pure rate of interest, also known as the risk-free rate of interest.

Capital market theory holds that once equilibrium is established, then it is maintained. In equilibrium, there is no pressure to change. Sharpe states that capital market theory asks about relationship between expected returns and risk for (1) portfolios; and (2) securities. The implicit question concerns the appropriate measure of risk for (1) a portfolio and (2) a security. The optimal combination of risky securities is the market portfolio, in which a security's percentage is its market value as compared to the total market value of risky securities.

Capital Market Line

The market portfolio includes only risky assets, and the actual return on the market portfolio is the weighted average of the expected returns of all risky securities. A line passes from the risk-free interest rate through the market portfolio on a return-risk graph. An individual preference curve, known as an indifference curve, determines where along this line, known as the Capital Market Line, the investor seeks to invest. A conservative investor might seek to lend money at the risk-free rate and invest the remaining funds in the market portfolio. An aggressive investor might borrow money at the risk-free rate and invest more funds than his or her initial endowment in the market portfolio.

To increase expected return along the Capital Market Line, the investor must accept more risk. Conversely, to reduce risk, an investor must give up expected return. Sharpe (1970) refers to the slope of the Capital Market Line as the price of risk reduction (in terms of decreased expected return). All efficient portfolios must lie along the capital market line

$$E(R_p) = R_F + r_e \sigma_p \qquad (8)$$

where $E(R_p)$ = expected portfolio return

r_e = price of risk reduction for efficient portfolios

R_F = pure (risk-free) rate of interest

σ_p = portfolio standard deviation

The Capital Market Line, equation (8), summarizes the simple (linear) relationship between expected return and risk of efficient portfolios. However, such a relationship does not hold for inefficient portfolios and individual securities.

Sharpe (1970) presents a very reader-friendly derivation of the Security Market Line (SML). Sharpe assumed that total funds were divided between the market portfolio, M, and security *i*. The investor is fully invested; hence

$$x_M + x_i = 1$$

The expected return of the portfolio is:

$$E(R_p) = x_i E(R_i) + x_M E(R_M) \tag{9}$$

and the corresponding variance of the portfolio is:

$$\sigma_p^2 = x_i^2 \sigma_i^2 + x_M^2 \sigma_M^2 + 2 x_i x_M \sigma_{iM}^2 \tag{10}$$

Assume that $E(R_i)$ is less than $E(R_m)$, as is its standard deviation. Sharpe presents this graphically as Figure 5.5.

Figure 5.5: Security Market Line and Two Securities Efficient Frontier

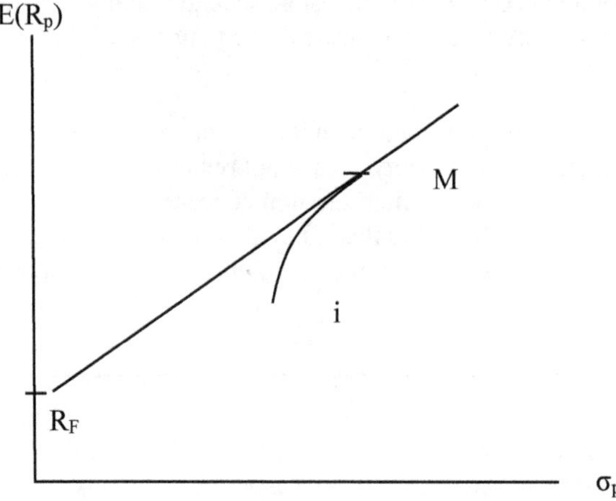

We know that the slope of the curve connecting the security *i* and market portfolio M depends on the relative weights, x_i and x_M, and the correlation coefficient, ρ_{iM}, between the security *i* and market portfolio *M*.

Rewrite $x_M = 1 - x_i$ and $\sigma_{iM} = \rho_{iM} \sigma_i \sigma_M$

$$\sigma_p = \sqrt{x_i^2 \sigma_i^2 + (1 - x_i)^2 \sigma_M^2 + 2 x_i (1 - x_i) \sigma_{iM}}$$

The portfolio risk's sensitivity to security i's weight is

$$\frac{\partial \sigma_p}{\partial x_i} = \frac{x_i(\sigma_i^2 + \sigma_M^2 - 2\sigma_{iM}) + \sigma_{iM} - \sigma_M^2}{\sigma_p} \tag{11}$$

Since portfolio expected return is linear to security *i*'s weight

$$E(R_p) = x_i E(R_i) + (1 - x_i) E(R_M)$$

The portfolio expected return's sensitivity to security *i*'s weight is

$$\frac{\partial E_p}{\partial x_i} = E(R_i) - E(R_M) \tag{12}$$

Use equation (11) and the chain rule to calculate portfolio expected return's sensitivity to risk:

$$\frac{\partial E_p}{\partial \sigma_p} = \frac{\partial E_p / \partial x_i}{\partial \sigma_p / \partial x_i} = \frac{E(R_i) - E(R_M)}{\dfrac{x_i(\sigma_i^2 + \sigma_M^2 - 2\sigma_{iM}) + \sigma_{iM} - \sigma_M^2}{\sigma_p}}$$

The above equation shows the trade-off between expected returns and portfolio standard deviations. At point *M*, the market portfolio, $x_i = 0$ and $\sigma_p = \sigma_M$, thus:

$$\frac{\partial E_p}{\partial \sigma_p} /_{x_i=0} = \frac{E(R_i) - E(R_M)}{(\sigma_{iM} - \sigma_M^2)/\sigma_M} = \frac{[E(R_i) - E(R_M)]\sigma_M}{\sigma_{iM} - \sigma_M^2} \tag{13}$$

In equilibrium, curve *iM* becomes tangent to the Capital Market Line, i.e., the slope of curve *iM* and the slope of the Capital Market Line must be equal. In other words, the trade-off of expected return and risk for the portfolio must be equal to the capital market trade-off. Otherwise, the market portfolio will not be an efficient portfolio.

Thus

$$\frac{[E(R_i) - E(R_M)]\sigma_M}{\sigma_{iM} - \sigma_M^2} = \frac{E(R_M) - R_F}{\sigma_M} \tag{14}$$

or

$$E(R_i) - R_F = \left[\frac{E(R_M) - R_F}{\sigma_M^2}\right]\sigma_{iM} \tag{15}$$

and

$$E(R_i) = R_F + [E(R_M) - R_F]\frac{\sigma_{iM}}{\sigma_M^2} \tag{16}$$

CAPM Beta

Sharpe (1970) discusses the stock beta as the slope of the firm's characteristic line, which is also the volatility of the security's return relative to changes in the market return.

The CAPM holds that the return to a security is a function of the security's beta.

$$R_{jt} = R_F + \beta_j[E(R_{Mt}) - R_F)] + e_{jt} \tag{17}$$

Where R_{jt} = expected security j return at time t

$E(R_{Mt})$ = expected return on the market at time t

R_F = risk-free rate

β_j = security beta, a random regression coefficient

e_{jt} = randomly distributed error term

Equation (17) defines the Security Market Line (SML), which describes the linear relationship between the security's return and its systematic risk, as measured by beta. Remember that we estimated stock betas in Chapter 3.

5.4 The Barra Model: A Fundamental Risk Model

The majority of institutional investors have been introduced to risk through the Barra risk model. In 1975, Barr Rosenberg and his associates introduced the Barra US Equity Model, often denoted USE1. We spend a great deal of time on the Barra USE1 and USE3 models because 70 of the 100 largest investment managers use the BARRA United States Equity Risk Models.[6] The Barra USE1 Model predicted risk, which required the evaluation of the firm's response to economic events, which were measured by the company's fundamentals, as discussed in Rudd and Clasing (1982).[7] The United States Equity Models have been updated and enhanced – USE2 in 1985, USE3 in 1997, and USE4 in 2011.

GEM

The first Barra Global Equity Risk Model (GEM) was introduced in 1989. The model was estimated via monthly cross-sectional regressions using countries, industries, and styles as explanatory factors, as described by Grinold, Rudd, and Stefek (1989). GEM2 was introduced in 2008 and published in Menchero, Morozov, and Shepard (2010). GEM2 incorporated several advances over the previous model, such as improved estimation techniques, higher-frequency observations, and the introduction of the World factor to place countries and industries on an equal footing. GEM3 was introduced in 2011. Given that Barra, now Morgan Stanley Capital International (MSCI) Barra, is the most successful commercial risk model and most widely used risk model, it seems appropriate for us to discuss its creation and structure.

Multiple Factor Model

Barr Rosenberg and Walt McKibben (1973) estimated the determinants of security betas and standard deviations. This estimation formed the basis of the Rosenberg extra-market component study (1974), in which security-specific risk could be modeled as a function of financial descriptors, or known financial characteristics, of the firm. The Rosenberg and Marathe (1979) paper developed the econometric methodology of investment returns analysis. Capital market equilibrium linked both first (mean) and second (variance) moments. Rosenberg and Marathe discussed econometric estimation techniques for second moments.[8] The total excess return for a multiple factor model, referred to as the MFM, in the Rosenberg methodology for security j, at time t, dropping the subscript t for time, can be written:

$$E(R_j) = \sum_{k=1}^{K} \beta_{jk}\tilde{f}_k + \tilde{e}_j \tag{18}$$

The non-factor, or asset-specific, return on security j, is the residual risk of the security, after removing the estimated impacts of the K factors. The term, f_k, is the rate of return on factor k. A single factor model, in which the market return is the only estimated factor, is obviously the basis of the Capital Asset Pricing Model. Accurate characterization of portfolio risk requires an accurate estimate of the covariance matrix of security returns. A relatively simple way to estimate this covariance matrix is to use the history of security returns to compute each variance, covariance, and security beta. The use of beta, the covariance of security and market index returns, is one method of estimating a reasonable cost of equity funds for firms.

However, the approximation obtained from simple models might not yield the best possible cost of equity. The simple, single index beta estimation approach suffers from two major drawbacks:

- Estimation of a stable beta requires a long history of return data. The company's beta might have been changed.
- It is subject to estimation error. Thus, one might expect a higher correlation between DuPont and Dow than between DuPont and IBM, given that DuPont and Dow are both chemical firms.

Taking this further, one can argue that firms with similar characteristics, such as firms in their line of business, should have returns that behave similarly. For example, DuPont and IBM will all have a common component in their returns because they would all be affected by news that affects the stock market, measured by their respective betas. The degree to which each of the three stocks responds to this stock market component depends on the sensitivity of each stock to the stock market component.

Additionally, one would expect DuPont and Dow to respond to news effecting the chemical industry, whereas IBM and Dell would respond to news effecting the computer industry. The effects of such news can be captured by the average returns of stocks in the chemical industry and the computer industry. One can account for industry effects in the following representation for returns:

$$\begin{aligned}\tilde{r}_{DD} &= E[\tilde{r}_{DD}] + \beta \cdot [\tilde{r}_M - E[\tilde{r}_M]] \\ &+ 1 \cdot [\tilde{r}_{CHEMICAL} - E[\tilde{r}_{CHEMICAL}]] + 0 \cdot [\tilde{r}_C - E[\tilde{r}_{DD}]] + \mu_{DD}\end{aligned} \quad (19)$$

where:

$\tilde{r}_{DD}$ = Dupont's (ticker symbol DD) realized return

$\tilde{r}_M$ = the realized average stock market return

$\tilde{r}_{CHEMICAL}$ = realized average return to chemical stocks

$\tilde{r}_C$ = the realized average return to computer stocks

E[.] = expectations

β_{DD} = DD's sensitivity to stock market returns

μ_{DD} = the effect of DD-specific news on DD returns

This equation simply states that DD's realized return consists of an expected component and an unexpected component. The unexpected component depends on any unexpected events that affect stock returns in general $[\tilde{r}_M - E[\tilde{r}_M]]$, any unexpected events that affect the chemical industry $[\tilde{r}_{CHEMICAL} - E[\tilde{r}_{CHEMICAL}]]$, and any unexpected events that affect DD alone (μ_{DD}). Thus, the sources of variation in DD's stock returns are variations in stock returns in general, variations in chemical industry returns, and any variations that are specific to DD. Moreover, DD and Dow returns are likely to move together because both are exposed to stock market risk and chemical industry risk. DD and IBM, on the other hand, are likely to move together to a lesser degree because the only common component in their returns is the market return.

Investors look at the variance of their total portfolios to provide a comprehensive assessment of risk. To calculate the variance of a portfolio, one needs to calculate the covariances of all the constituent components. Without the framework of a multiple-factor model, estimating the covariance of each asset with every other asset is computationally burdensome and subject to significant estimation errors. The Barra MFM simplifies these calculations dramatically, replacing individual company profiles with categories defined by common characteristics (factors). The specific risk is assumed to be uncorrelated among the assets and only the factor variances and covariances are calculated during model estimation.

The multiple-factor risk model significantly reduces the number of calculations inherent in covariance analyses. For example, in the US Equity Model (USE3), 65 factors capture the risk characteristics of equities. This reduces the number of covariance and variance calculations; moreover, since there are fewer parameters to determine, they can be estimated with greater precision. The BARRA risk management system begins with the MFM equation:

$$\tilde{r}_i = X\tilde{f} + \tilde{u} \tag{20}$$

where:

$\tilde{r}_i$ = excess return on asset *i*

X = exposure coefficient on the factor

$\tilde{f}$ = factor return

$\tilde{u}$ = specific return

Substituting this relation in the basic equation, we find that:

$$\text{Risk} = \text{Var}(\tilde{r}_j) \tag{21}$$

$$= \text{Var}(X\tilde{f} + \tilde{u}) \tag{22}$$

Using the matrix algebra formula for variance, the risk equation becomes:

$$\text{Risk} = XFX^T + \Delta \tag{23}$$

where:

X = exposure matrix of companies upon factors

F = covariance matrix of factors

X^T = transpose of X matrix

Δ = diagonal matrix of specific risk variances

This is the basic equation that defines the matrix calculations used in risk analysis in the BARRA equity models.[9] Investment managers seek to maximize portfolio return for a given level of risk. For many managers, risk is measured by the BARRA risk model.[10]

Historical Beta

The most frequent approach to predict risk is to use historical price behavior in the estimation of beta. Beta was defined as the sensitivity of the expected excess rate of return on the stock to the expected excess rate of return on the market portfolio. A major assumption has to be made to enable average (realized) rates of return to be used in place of expected rates of return, which, in turn, permits one to use the slope of regression line (estimated from realized data) to form the basis for a prediction of beta.

If this assumption, which essentially states that the future is going to be similar to the "average past," is made, then the estimation of historical beta proceeds as follows. When more data points are used, the accuracy of the estimation procedure is improved, provided the relationship being estimated does not change. Usually, the relationship does change; therefore, a small number of most recent data points is preferred so that dated information will not obscure the current relationship. It is usually accepted that a happy medium is achieved by using 60 monthly returns.[11] The security return series is then regressed against the market portfolio return series. This provides an estimate of beta (which is equivalent to the slope of the characteristic line) and the residual variance.

It can be shown that if the regression equation is properly specified and certain other conditions are fulfilled, then the beta obtained is an optimal estimate (actually, minimum-variance, unbiased) of the true historical beta averaged over past periods. However, this does not imply that the historical beta is a good predictor of future beta. The beta obtained by this method is an estimate of the true historical beta and can be obscured by measurement error. Rudd and Clasing (1982) discussed beta prediction with respect to the use of historic price information. Three possible prediction methods for beta were suggested. These are:

1. *Naïve*: $\hat{\beta}_j = 1.0$ for all securities (i.e., every security has the average beta).
2. *Historical*: $\hat{\beta}_j = H\hat{\beta}_j$, the historical beta obtained as the coefficient from an ordinary least squares regression.
3. *Bayesian adjusted beta*: $\hat{\beta}_j = 1.0 + BA(H\hat{\beta}_j - 1)$, where the historical betas are adjusted toward the mean value of 1.0.

In each case, the prediction of residual risk is obtained by subtracting the systematic variance $(\hat{\beta}_j^2 V_M)$ from the total variance of the security. The predicted residual risk can be different from the residual variance obtained directly from the regression.

The historical beta estimate will be an unbiased predictor of the future value of beta, provided that the expected change between the true value of beta averaged over the past periods and its value in the future is zero. If this expected change is not zero, then the historical beta estimate will be misleading (biased). The Standard & Poor's (S&P) Corporation estimates its beta using five years of monthly returns. Industry estimates betas with years of monthly observation.

The empirical evidence regarding the construction of the Barra risk models comes from several well-known publications by Barr Rosenberg. Barr Rosenberg and Walt McKibben (1973) estimated the determinants of security betas and standard deviations. This estimation formed the basis of the Rosenberg extra-market component study (1974), in which security systematic and specific risk could be modeled as a function of financial descriptors, or known financial characteristics of the firm. Rosenberg and McKibben found that the financial characteristics that were statistically associated with beta during the 1954–1970 period were:

1. Latest annual proportional change in earnings per share
2. Liquidity, as measured by the quick ratio
3. Leverage, as measured by the senior debt-to-total assets ratio
4. Growth, as measured by the growth in earnings per share
5. Book-to-Price ratio
6. Historic beta
7. Logarithm of stock price
8. Standard deviation of earnings per share growth
9. Gross plant per dollar of total assets
10. Share turnover

Rosenberg and McKibben used 32 variables and a 578-firm sample to estimate the determinants of betas and standard deviations. For betas, Rosenberg and McKibben found that the positive and statistically significant determinants of beta were the standard deviation of eps growth, share turnover, the price-to-book multiple, and the historic beta.[12] The statistically significant determinants of the security systematic risk became the basis of the Barra E1 Model risk indexes.

USE1

The fundamental prediction method of Barra starts by describing the company, see Rudd and Clasing (1982). The Barra USE1 Model estimated "descriptors," which are ratios that describe the fundamental condition of the company. These descriptors are grouped into six categories to indicate distinct sources of risk. In each case, the category is named so that a higher value is indicative of greater risk.

1. *Market variability*. This category is designed to capture the company as perceived by the market. If the market was completely efficient, then all information on the state of the company would be reflected in the stock price. Here the historical prices and other market variables are used in an

attempt to reconstruct the state of the company. The descriptors include historical measures of beta and residual risk, nonlinear functions of them, and various liquidity measures.

2. *Earnings variability.* This category refers to the unpredictable variation in earnings over time, so descriptors such as the variability of earnings per share and the variability of cash flow are included.
3. *Low valuation and unsuccess.* If investors are optimistic about future prospects and the company has been successful in the past (measured by a low book-to-price ratio and growth in per share earnings), then the implication is that the firm is sound and that future risk is likely to be lower. Conversely, an unsuccessful and lowly valued company is more risky than the average stock.
4. *Immaturity and smallness.* A small, young firm is likely to be more risky than a large, mature firm.
5. *Growth orientation.* To the extent that a company attempts to provide returns to stockholders by an aggressive growth strategy requiring the initiation of new projects with uncertain cash flows rather than the more stable cash flows of existing operations, the company is likely to be more risky. Thus, the growth in total assets, payout and dividend policy, and the earnings/price ratio are used to capture the growth characteristics of the company.
6. *Financial risk.* The more highly levered the financial structure, the greater is the risk to common stockholders.

Finally, the industry in which the company operates is another important source of information. Certain industries, simply because of the nature of their business, are exposed to greater (or lesser) levels of risk (e.g., compare airlines versus gold stocks). Rosenberg and Marathe used indicator (dummy) variables for 39 industry groups as the method of introducing industry effects.

The Barra USE1 Model predicted risk, which required the evaluation of the firm's response to economic events, which were measured by the company's fundamentals, is discussed in Rudd and Clasing (1982). There are three major steps. First, for the time period during which the model is to be fitted, obtain common stock returns and company annual reports (for instance, from the COMPUSTAT database).[13] In order to make comparisons across firms meaningful, the descriptors must be normalized so that there is a common origin and unit of measurement.

The normalization takes the following form. First, the "raw" descriptor values for each company are computed. Next, the capitalization weighted value of each descriptor for all the securities in the S&P 500 is computed and then subtracted from each raw descriptor. The transformed descriptors now have the property that the capitalization weighted value for the S&P 500 stocks is zero. Furthermore, the standard deviation of each descriptor is calculated within a universe of large companies (defined as having a capitalization of $50 million or more). The descriptor is now further transformed by setting the value + 1 to be one standard deviation above the S&P 500 mean (i.e., one unit of length corresponds to one standard deviation). Rudd and Clasing (1982) write:

$$ND = (RD - RD\,[S\&P])/STDEV[RD] \qquad (24)$$

where

ND = the normalized descriptor value

RD = the raw descriptor value as computer from the data

RD[S&P] = the raw descriptor value for the (capitalization-weighted) S&P 500

Each company is identified by a series of descriptors that indicate its fundamental position. If a descriptor value is zero, then the company is "typical" of the S&P 500 (for this characteristic) because the S&P 500 and the company both have the same raw value. Conversely, if the descriptor value is nonzero, then the company is atypical of the S&P 500, and this information can be used to adjust the prior prediction in order to obtain a better posterior prediction of risk.

If the company is completely typical of market (i.e., the descriptors other than historical beta are all zero), then there is no further adjustment to the Bayesian-adjusted historical beta. If the company is atypical, then not all the descriptors (other than historical beta) will be zero. The Barra risk model estimates the company's exposure to each of the common factors and the prediction of the residual risk components. The first task is to form summary measures or indices of risk to describe all aspects of the company's investment risk. These are obtained by forming the weighted average of the descriptor values in each of the six categories introduced previously in this section where the weights are the estimated coefficients from the prediction rule for systematic or residual risk. This provides six summary measures of risk – the risk indices – for each company. Again, these indices are normalized so that the S&P 500 has a value of zero on each index and a value of one corresponds to one standard deviation among all companies with capitalization of $50 million or more.

The prediction of residual risk is now found by performing a regression on the cross section of all security residual returns as the dependent variable where the independent variables are the risk indices.[14] The form of the regression, in a given month, is shown in equation (25):

$$r_i - \hat{\beta}_i r_M = c_1 RI_{1i} + \ldots + c_6 RI_{6i} + c_7 IND_{1i} + \ldots + c_{45} IND_{39i} + u_i \quad (25)$$

where

r_i = the excess return on security

$\hat{\beta}_i$ = the predicted beta

r_M = the excess return on the market portfolio

so that $r_i - \hat{\beta}_i r_M$ is the residual return on security *i*; $RI_{1i}, \ldots, RI_{6i}$, are the six risk indices for security *i*, $IND_{1i}, \ldots, IND_{39,i}$ are the dummy variables for the 39 industry groups; u_i, is the specific return for security *i*; and $c_1, \ldots, c_{45}$ are the 45 coefficients to be estimated.[15]

The entire risk of the security arises from two sources: the systematic or factor risk ($b_j^2\text{Var}[f]$), and the nonfactor risk (σ_j^2). In this case, however, the nonfactor risk is completely specific risk since no risk arises from interactions with other securities. In other words, under these assumptions the single factor, *f*, is responsible for the only commonality among security returns. Thus, the random return component that is not related to the factor must be specific to the individual security, *j*.

If we form a portfolio, *P*, with weights $h_{P1}, h_{P2}, \ldots, h_{PN}$, from *N* stocks, then the random excess return on the portfolio is given by:

$$R_P = \Sigma h_{Pj} r_j = \Sigma h_{Pj} b_j f + \Sigma h_{Pj} u_j = b_P f + \Sigma h_{Pj} u_j, \quad (26)$$

where $b_P = \Sigma h_{Pj} r_j$. The mean return and variance are:

$E[r_P] = a_P + b_P\, E[f]$

where $a_P = \Sigma h_{Pj} a_j$

$$\text{Var}[r_P] = b_p^2\, \text{Var}[f] + \Sigma h_P^2 \sigma_j^2 \quad (27)$$

where we have used the fact that the security-specific risk is *specific*, i.e., independent across securities and independent of the factor return. The regression coefficient of an individual stock's rate of return onto the market, or beta, is given by:

$\beta_j = \text{Cov}[r_j, r_M] / \text{Var}[r_M]$

$= \text{Cov}[b_j f + u_j, f + \Sigma h_{Mk} u_k] / \text{Var}[r_M]$

$= (b_j \text{Var}[f] + h_{Mj}\sigma_j^2) / \text{Var}[r_M]$

$$= (b_j \text{Var}[f] + h_{Mj}\sigma_j^2) / (\text{Var}[f] + \Sigma h_{Mj}^2 \sigma_j^2) \quad (28)$$

so that:

$\beta_P = (b_P \text{Var}[f] + \Sigma h_{Mj} h_{Pj} \sigma_j^2) / (\text{Var}[f] + \Sigma h_{Mj}^2 \sigma_j^2)$.[16]

USE3 and USE4

The domestic BARRA E3 (USE3, or sometimes denoted US-E3) model, with some 15 years of research and evolution, uses 13 sources of factor, or systematic, exposures. The sources of extra-market factor exposures are volatility, momentum, size, size non-linearity, trading activity, growth, earnings yield, value, earnings variation, leverage, currency sensitivity, dividend yield, and non-estimation universe indicator.[17] In November 2011, Barra introduced USE4(L), the Barra US Equity Model Long-Term Version, which featured 12 factors, including beta, momentum, size, earnings yield, and growth. An asset manager could prefer to use a price momentum or earnings growth tilt in the portfolio, and in this case, only the momentum, growth, earnings yield, and size exposures might (should) be nonzero.[18]

Menchero and Nagy (2015) used the USE4 and GEM3 model to study the effectiveness of portfolio construction emphasizing price momentum and earnings yield factor in United States and Global equity portfolios.[19] The GEM3 and USE4 risk factors are described in Table 5.1.

Table 5.1: USE4 Descriptor Definitions

Descriptor	Definition
Beta	Historic beta
Non-Linear Beta	Non-linear historic beta
Residual Volatility	Composed of variables explaining returns of high-volatility stocks not associated with the estimated beta factor including the daily standard deviation
Momentum	Composed of a cumulative six- to twelve-month relative strength variable
Size	Log of the security market capitalization
Non-Linear Size	The cube of the log of the security market capitalization
Liquidity	Annualized share turnover of the past twelve months, quarter, and month
Growth	Growth in total assets, five-year growth in earnings per share, sales growth, and analyst-predicted sales growth
Earnings Yield	Consensus analyst-predicted earnings to price and the historic earnings to price ratios, and cash earnings-to-price ratio
Book-to-Price	Book to price ratio
Leverage	The debt-to-assets ratio and market and book value leverage
Dividend Yield	Dividend-to-price ratio

The definitions of the descriptors underlie the risk indices in USE4 (L, denoting long-term version). The method of combining these descriptors into risk indices is proprietary to Barra. The USE4 model starts in June 1995 and uses the United States component of MSCI All Country World Investable Market Index (IMI).

The GEM3 risk factor indices are identical to the USE4 factor risk indices with the omission of Non-Linear Beta.

BARRA Applications

The authors have used the BARRA system for over 20 years. There is no question that BARRA has been state-of-the-art in realistic portfolio simulations for well over 30 years, beginning in 1979, see Rudd and Rosenberg (1979). Guerard and Mark (2003) reported highly statistically significant asset selection with an earnings forecasting model (CTEF) that continues to produce positive and statistically significant specific returns, for the 2003–2016 time period, see Guerard and Mark (2018). See Table 5.2 for adapted results from Guerard and Mark (2003) and Table 5.3 for adapted results from Guerard and Mark (2018), where CTEF dominates the composite regression models for US stock selection introduced in Chapter 4.

Table 5.2: Components of the Composite Earnings Forecasting Variable, 1990–2001

Russell 3000 Universe

R3000 Earnings Analysis	Total Active	T-stat	Asset Selection	T-stat	Risk Index	T-stat
FEP1	2.14	1.61	-1.18	-1.17	4.20	4.42
FEP2	1.21	0.91	-1.43	-1.35	3.33	3.35
RV1	0.76	0.69	0.34	0.42	0.92	1.46
RV2	1.40	1.37	1.09	1.31	0.81	1.42
BR1	2.59	2.83	1.85	2.43	1.08	2.15
BR2	2.43	2.36	1.51	1.75	1.09	2.04
CTEF	2.87	2.81	2.07	2.66	1.19	1.70

The CTEF variable produced statistically significant total active returns and asset selection in Table 5.2, adapted from Guerard and Mark (2003).

The ranked CTEF (RCTEF) variable produced statistically significant total active returns and asset selection in Table 5.3, adapted from Guerard and Mark (2018), for the 2003–2016 time period. Note that as we expand the CTEF Model to include the eight variables of the DPOS Markowitz Model, REG9_CTEF, and add the PM121 Price Momentum variable, to create REG10, discussed in Chapter 4, that Total Active Returns and Specific returns decrease in the US portfolios.

Table 5.3: Axioma Attribution Verification of US Portfolios, 2003–2016

Russell 3000 Index Constituents						Risk Attribution			
Portfolio	Port. Total Return	Bench. Total Return	Active Total Return	Factors Effect	Risk Factors Effect T-Stat	Stock Specific Effect	Risk Stock Specific Effect T-Stat	Total Effect
RCTEF_MV4	**15.69**	**7.92**	**7.77**	**5.02**	**4.91**	**2.75**	**2.62**	**7.77**
REG10_MV4	**13.98**	**7.92**	**6.06**	**6.22**	**5.31**	**-0.16**	**0.22**	**6.06**
REG8_MV4	**13.76**	**7.92**	**5.83**	**5.32**	**3.83**	**0.52**	**1.27**	**5.83**
REG9_CTEF_MV4	**13.68**	**7.92**	**5.76**	**6.28**	**4.76**	**-0.52**	**0.25**	**5.76**

Miller, Xu, and Guerard (2014) documented the effectiveness of CTEF using USE3 for the 1979–2009 time period, see Table 5.4, using the S&P 500 Index as its benchmark; 125 basis points, each way, of transactions costs; and a maximum stock weight of 4 percent in the portfolio. The BARRA system offers the longest backtest period available in commercially available optimization and simulation systems, to the best of the authors' knowledge.

Table 5.4: USER Efficient Frontier, Barra Risk Model and Optimizer, 1979–2009

Risk Aversion Level (RAL)	Total Managed Return	Portfolio STD	Total Active Return	Asset Selection	IR (t)	Tracking Error	N	Transactions Costs
0.010	13.69	17.65	7.79 (2.42)	6.75 (3.27)	0.76	10.05	67	2.74
0.020	12.24	17.30	6.34 (2.19)	5.24 (2.77)	0.69	9.13	80	2.73
0.025	12.11	17.34	6.21 (2.16)	5.26 (2.80)	0.69	9.01	84	2.73
0.050	10.05	16.77	4.15 (1.78)	4.02 (2.51)	0.56	7.51	104	2.67
0.090	9.51	16.33	3.61 (1.78)	4.01 (2.75)	0.56	6.46	124	2.65
0.150	8.82	16.01	2.92 (1.65)	3.70 (2.77)	0.52	5.64	147	2.62
0.200	8.14	15.77	2.24 (1.39)	3.26 (2.63)	0.44	5.16	163	2.60

Several results should be noted from the reported simulations of Table 5.4 and its thirty-year backtest.

1. Over the long-run, for the USER Model, and many other statistically significant asset selection models in the US, the use of a tracking error of less than 5 percent will most likely not produce statistically significant Total Active Returns
2. The use of the BARRA optimizer and USE3 Risk model with the USER Model as a tilt variable produces statistically significant Total Managed Returns, Total active Returns, and Asset Selection at the 5 percent level, the -statistic exceeding 1.96, as long as the Risk Aversion Level, RAL, exceeds 0.025 and the optimal average number of stocks, N, is less than 84
3. One must apply the model every year and target a 7–8 percent tracking error.

5.5 APT and Statistical Risk Models: Constructing Mean-Variance Efficient Portfolios

The origin of modern finance in this context must be traced to the work of Markowitz (1952–1959). The conceptual framework is based on the work of von Neuman-Morgenstern (1944) who pioneered the view that choice under uncertainty can be based on expected utility. As initially formulated by Markowitz, this involved the maximization of portfolio returns given a variance constraint. In practice, the standard deviation was substituted for the variance to eliminate dependence on the units of measurement, e.g. dollars versus thousands of dollars. At the optimum, the Lagrange multiplier, λ, equals what came to be known as the Sharpe ratio.

The interpretation of λ in other areas of economics is that of a "shadow price" and measures the extent to which the function to be maximized (expected portfolio returns) will change by relaxing the constraint (risk) at the optimum. Thus, it can be interpreted as the marginal return on risk, or the reward (or price) of risk at the optimum.

This has led to the widespread belief that the portfolio manager should seek to maximize the portfolio geometric mean (GM) and Sharpe ratio (ShR) as put forth in Latane (1959) and Markowitz (1959 and 1976).[20] However, as formulated in Markowitz, the portfolio manager seeks to identify the efficient frontier, the point at which the portfolio return is maximized for a given level of risk, or, equivalently, portfolio risk is minimized for a given level of portfolio return.

The Markowitz framework measures risk as the portfolio standard deviation, a measure of dispersion or total risk. Investors seek to minimize risk, as measured by the covariance matrix in the Markowitz framework, by holding constant expected returns. Elton et al. (2007) proposed an equivalent formulation of the traditional Markowitz mean-variance problem as a maximization problem, i.e. to maximize

$$\theta = \frac{E(R_p) - R_F}{\sigma_p} \tag{29}$$

where R_F is the risk-free rate (typically measured by the 90-day Treasury bill rate). A little reflection, however, will show that it is not. In fact, these are two conceptually separate problems.

APT Approach

John Blin, Steve Bender, and John Guerard (1997) and Guerard (2012) demonstrated the effectiveness of the arbitrage pricing theory model (APT). John Blin and Steve Bender developed Advanced Portfolio Technologies, "their APT", to commercialize their portfolio construction and management research. The Blin and Bender APT was acquired by Sungard APT, and is now FIS APT. Let us review the APT approach to portfolio construction. The estimation of security weights, *w*, in a portfolio is the primary calculation of Markowitz's portfolio management approach. The issue of security weights will be now considered from a different perspective. The security weight is the proportion of the portfolio's market value invested in the individual security.

$$w_s = \frac{MV_s}{MV_p} \tag{30}$$

where w_s = portfolio weight in security s, MV_s = value of security *s* within the portfolio and MV_p = the total market value of portfolio.

The active weight of the security is calculated by subtracting the security weight in the (index) benchmark, *b,* from the security weight in the portfolio, *p*

$$w_{sp} - w_{sb} \tag{31}$$

Blin and Bender's Advanced Portfolio Technologies Analytics Guide (2005) built upon the mathematical foundations of their APT system, published in Blin, Bender, and Guerard (1997). The following analysis draws upon the APT analytics. Volatility can be broken down into systematic and specific risk:

$$\sigma_p^2 = \sigma_{\beta p}^2 + \sigma_{\varepsilon p}^2 \tag{32}$$

where σ_p = Total Portfolio Volatility, $\sigma_{\beta p}$ = Systematic Portfolio Volatility and $\sigma_{\varepsilon p}$ = Specific Portfolio Volatility. Blin and Bender created a multi-factor risk model within their APT risk model based on forecast volatility.

$$\sigma_p = \sqrt{52 * (\sum_{c=1}^{C} \sum_{i=1}^{S} w_i \beta_{i,c})^2 + 52 * \sum_{i=1}^{S} w_i^2 \varepsilon_i^2} \tag{33}$$

Where σ_p = forecast volatility of annual portfolio return

C = number of statistical components in the risk model

w_i = Portfolio weight in security *i*

$\beta_{i,c}$ = loading (beta) of security *i* on risk component *c*

$\varepsilon_{i,week}$ = weekly specific volatility of security *i*

The multiplier 52 is the annualization factor since there are 52 weeks per year.

Tracking error is a measure of volatility applied to the active return of funds (or portfolio strategies) against a benchmark, which is often an index. Portfolio tracking error is defined as the standard deviation of the portfolio return less the benchmark return over one year.

$$\sigma_{te} = \sqrt{E((r_p - r_b) - E(r_p - r_b))^2} \tag{34}$$

Where σ_{te} = annualized tracking error
r_p = actual portfolio annual return
r_b = actual benchmark annual return

Systematic tracking error of a portfolio is a forecast of the portfolio's active annual return as a function of the securities' returns associated with APT risk model components. Portfolio-specific tracking error can be written as a forecast of the annual portfolio active return associated with each security's specific behavior.

$$\sigma_{\varepsilon te} = \sqrt{52 \sum_{i=1}^{N_s} (w_{i,p} - w_{i,b})^2 \varepsilon_{i,week}^2} \tag{35}$$

The APT calculated portfolio tracking error versus a benchmark as:

$$\sigma_{p-b} = \sqrt{52(w_p - w_b)^T (b^T b + \varepsilon^T \varepsilon)(w_p - w_b)}, \tag{36}$$

Where σ_{p-b}= forecast tracking error

b= A $(N_c \times N_s)$ matrix of component loadings; N_c components in the model and N_s securities in the portfolio,

ε= A diagonal matrix $(N_c \times N_s)$ of the specific loadings,

$w_p - w_b$ = The (N_s -dimensional) vector security active weights.

$$\sigma_{p-b} = \sqrt{52[(\sum_{c=1}^{N_c} \sum_{s=1}^{N_s} (w_{sp} - w_{sb}) b_{sc})^2 + \sum_{s=1}^{N_s} (w_{sp} - w_{sb})^2 \varepsilon_s^2]} \tag{37}$$

Where $w_{sp} - w_{sb}$ = portfolio active weights in security s

b_{sc} = the loading of security s on component c

ε_s = weekly specific volatility of security s

The marginal tracking error is a measure of the sensitivity of the tracking error of an active portfolio to changes in active weight of a specific security.

$$\partial_s[\sigma_{p-b}] = \frac{\partial \sigma_{p-b}}{\partial w_{s,p-b}} \tag{38}$$

Where $w_{s,p-b}$ = active portfolio weights in security s.

The APT calculated contribution-to-risk of a security is

$$\Delta[\sigma_p] = \frac{52(b^T b + s^T s)w}{\sqrt{52 w^T (b^T b + s^T s) w}} \tag{39}$$

where $\Delta[\sigma_p]$ = A $(N_s - dimensional)$ vector of contribution to volatility for securities in the portfolio with N_s securities.

The portfolio Value-at-Risk (VaR) is a measure of the distribution of expected outcomes. If we are concerned with a 95 percent confidence level, α, and a 30-day time horizon, then the 95 percent, 30-day VaR of the portfolio is the minimum that we would expect to lose, which is characterized by

$$P[V_T < V_0 - v(\alpha, T)] = 1 - \alpha \tag{40}$$

where $v(\alpha, T)$ = VaR at the confidence interval α for time T

V_T = portfolio value at time T

α = required confidence level

If the portfolio returns are normally distributed,

$$v_s(\alpha, T) = V_0 - E(V_T) - V_0 \phi^{-1}(\alpha) \sigma_p^T \tag{41}$$

where $v_s(\alpha, T)$ = Gaussian VaR,

σ_p^T = forecast portfolio volatility

$\phi^{-1}(\alpha)$ = inverse cumulative normal distributed function

The total VaR is

$$v_c(\alpha, T) = V_0 \sqrt{\frac{\alpha}{1-\alpha}(\sigma_{\beta p}^T + (\phi^{-1}(\alpha)\sigma_{sp}^T)^2)} \tag{42}$$

The Tracking Error at Risk (TaR) is a measure of portfolio risk estimating the magnitude that the portfolio return might deviate from the benchmark return over time. The maximum deviation of a portfolio return over the time horizon *T* at given confidence level α is

$$v_{p-b}(\alpha, T) = V_0 \sqrt{(\frac{1}{\sqrt{1-\alpha}}\sigma_{\beta,p-b}^T)^2 + \phi^{-1}(\frac{1+\alpha}{2})\sigma_{s,p-b}^2} \tag{43}$$

Applying the Blin and Blender APT Model

Blin, Bender, and Guerard (1997) used an estimated 20-factor beta model of covariances based on 3.5 years of weekly stock returns data. The Blin and Bender Arbitrage Pricing Theory (APT) model followed the Ross factor modeling theory, but Blin and Bender estimated betas from at least 20 orthogonal factors. Empirical support is reported in Guerard, Chettiappan, and Xu (2010); Guerard (2012); and Guerard. See Krauklis and Kumar (2012) for the application of mean-variance, enhanced index tracking, and tracking error at risk optimization techniques.

It is well known that as one raises the portfolio lambda, the expected return of a portfolio rises and the number of securities in the optimal portfolios fall, see Blin, Bender, and Guerard (1997). Lambda, a measure of risk-aversion, the inverse of the risk-aversion acceptance level of the Barra system, is a decision variable to determine the optimal number of securities in a portfolio. Guerard, Chettiappan, and Xu (2010) reported a lambda of 200 maximized the geometric mean in non-US growth portfolios and Guerard, Krauklis, and Kumar (2012) reported that the lambda of 200 is a necessary lambda with MV, EAW3, and EAW4 portfolio construction models for the USER data to create concentrated portfolios with fewer than 100 securities.[21]A lambda of 200 implies optimal portfolios of 99.7 (100) stocks with mean-variance, MV, whereas Mean-Variance Tracking Error at Risk (MVTaR) requires only 77.8 (78) stocks.[22]

The Blin and Bender Tracking at Error (TaR) optimization procedure allows a manager to use fewer stocks in his or her portfolios than a traditional mean-variance optimization technique manager for a given lambda. In spite of the Markowitz Mean-Variance (M59) portfolio construction and management analysis being six decades old, it does very well in maximizing the Sharpe Ratio, Geometric Mean, and Information Ratio relative to newer approaches. See Table 5.5 for an adapted version of the Markowitz Mean-Variance and MVTaR simulations of Guerard, Markowitz, and Xu (2015).

Table 5.5: Efficient Frontier of the Global Selection Model, 1999–2011

Earnings Model or Component	Mean-Variance Methodology	Lambda	Annualized Return	Sharpe Ratio	Information Ratio	Tracking Error
GLER	M59	1000	15.84	0.590	0.78	13.11
		500	16.34	0.590	0.82	12.08
		200	16.37	0.610	0.85	12.68
		100	15.90	0.580	0.81	12.66
		5	10.11	0.440	0.51	8.81
Benchmark			5.59	0.240		
GLER	MVTaR	1000	16.10	0.660	0.94	11.18
		500	15.91	0.651	0.90	11.44
		200	16.09	0.691	0.97	10.83
		100	14.18	0.591	0.77	11.23
		5	8.51	0.344	0.33	8.75
GLER	EAWTaR2	1000	14.80	0.600	0.94	11.07
		500	14.30	0.590	0.80	10.87
		200	14.15	0.600	0.85	10.04
		100	13.49	0.570	0.80	9.84
		5	10.77	0.440	0.43	12.18
Constraints:						
1. Turnover is 8% (buys) monthly;						
2. Long-only holdings; and						
3. Threshold positions of 35 basis points.						

A lambda of 200, an aggressive risk-tolerance parameter, is sufficient in US and global markets to maximize the Geometric Mean, Sharpe Ratio, and Information Ratio (IR) for the MCM variables. The lambda of 200 MVTaR GLER simulation outperforms the M59 (based on Markowitz's seminal Portfolio Selection, one of the two sleeping partners of the authors) on only the IR basis; 0.97 versus 0.85). MVTaR does not "beat the tar" out of the traditional Markowitz Mean-Variance model, but it does produce lower tracking errors and higher IRs. A lambda of 5 is reported to illustrate an index-enhanced portfolio strategy.

Applying the APT Model with a Corporate Exports Constraint

In Chapter 2 we introduced corporate exports. We will illustrate in this section the effectiveness of creating portfolios with a CTEF tilt as we did with Guerard and Mark (2003), but additionally constraining the portfolio to have very high levels of corporate exports. Let us compare US and global portfolios during the December 2002–May 2015 time period. The global security market covered by analysts in the I/B/E/S database contained approximately 16,000 securities in October 2014. A single investment manager probably cannot invest in all of these securities with equal effectiveness. An investment manager can invest in Russell 3000, US stocks, and MSCI-ACWI index, the MSCI global stock index, constituents. If we created a corresponding US MVTaR portfolio with a CTEF tilt, then the active returns would be 6.13 percent, highly statistically significant with a t-statistic of 2.35, composed of specific returns of 3.61 percent (t=2.02), and factor returns of 2.51 percent (t=1.29). The factor returns are not statistically significant in the US CTEF tilt, although the medium-term momentum exposure produces a 0.77 percent return and a highly statistically significant t-statistic of 2.23. See Figure 2 in Guerard, Markowitz, Xu, and Wang (2018). If we created a corresponding global MVTaR portfolio with a CTEF tilt, then the active returns would be 14.99 percent, highly statistically significant with a t-statistic of 4.93, composed of specific returns of 9.54 percent (t=4.42). The global CTEF portfolio dominates the US CTEF portfolio in Geometric Means, Information Ratios, and Sharpe Ratios, by approximately a 2:1 ratio. Global markets are more inefficient markets than the US market.

Table 5.6 Axioma Portfolio Attributes of Simulations, December 2002–May 2015

Optimization Technique	Universe	Model Tilt	Constraint	Active Returns (t)	Specific Returns (t)
MVTaR	MSCI-ACWI	CTEF		14.99 (4.93)	9.54 (4.42)
MVTaR	MSCI-ACWI	CTEF	CE	15.80 (5.36)	7.64 (4.10)
MVTaR	Russell 3000	CTEF		6.13 (2.35)	3.61 (2.02)
MVTaR	Russell 3000	CTEF	CE	6.45 (2.21)	4.81 (2.55)

If we created a corresponding US MVTaR portfolio with a CTEF tilt and CE constraint, then the active returns would be 6.45 percent, highly statistically significant with a t-statistic of 2.21, composed of specific returns of 4.81 percent (t=2.55). If we created a Global MVTaR portfolio using a CTEF tilt and constrained corporate exports to be high (exceeding 80), then the global active returns would rise to 15.80 percent, highly statistically significant with a t-statistic of 5.36. Specific returns would fall to 7.64 (t=4.10), but are still highly statistically significant. Corporate exports enhance portfolio returns in US and global portfolios. You can see the relevance of corporate exports.

The mean-variance and equal active weighting tracking error at risk optimization techniques with a CTEF tilt produce portfolios statistically significant Geometric Mean and Information Ratio maximization. The Geometric Means and Information Ratios rise with the imposition of a corporate exports constraint, as was reported in Guerard, Markowitz, and Xu (2014). Firms that pay higher dividends, buy back more stock, and issue less debt produce higher stock returns for their stockholders. Global stock portfolios dominate US stock portfolios in Geometric Mean, Information Ratios, and Sharpe Ratios.

5.6 The Axioma Risk Model: Fundamental and Statistical Risk Models

The Axioma Robust Risk Model[23] is a multiple-factor risk model in the tradition of the Barra model and equation (20). Axioma offers both US and World fundamental and statistical risk models. The Axioma Risk Models use several statistical techniques to efficiently estimate factors. The ordinary least squares residuals (OLS) are not homoscedastic; that is, when one minimizes the sum of the squared residuals to estimate factors using OLS, one finds that large assets exhibit lower volatility than smaller assets. A constant variance of returns is not found. Axioma uses a weighted least squares (WLS) regression, which scales the asset residual by the square root of the asset market capitalization (to serve as a proxy for the inverse of the residual variance).

Robust regression using the Huber M Estimator addresses the issue and problem of outliers. (Asymptotic) Principal components analysis (PCA) is used to estimate the statistical risk factors. A subset of assets is used to estimate the factors and the exposures and factor returns are applied to other assets. In 2011, McKinley Capital Management, LLC (MCM) initiated a "Horse Race" testing procedure to test if all optimizers were created equal. They are not. In 2011, APT and Axioma were the winners among several (4–5) optimization systems using the MCM US "Public Models" CTEF and USER. Both APT and Axioma optimization systems produced highly statistically significant asset selection as shown in Tables 5.7 and 5.8 (the BARRA attribution system was used as a judge).

Table 5.7: APT and Axioma Statistical Risk Models, US Stocks, 1998-2009.

Assumptions and Constraints

Universe: All US stocks in the WRDS Database
Time Period: Jan1998 to
Dec2009

Position Constraint :

Minimum Position =0.0

Maximum Position=4.0
Position Threshold Holding = 0.35%,
Turnover =8% Monthly; and

Transaction Cost: 1.25% each way

Risk Model :

Axioma - Axioma Statistical Model

APT - APT Statistical Model

Risk Setting:

Axioma - Tracking error target of 8%

APT - Lambda of 200

Model : Regression Weighted EP BP CP SP REP RBP RCP RSP CTEF PM

BenchMark: R3000G

	SR	IR	TE	SD	AR	Asset Selection	T-Stat
Axioma	0.177	0.69	7.78	21.92	7.06	6.00	5.09
APT	0.117	0.40	10.29	23.02	5.87	6.77	3.92
R3000G	-0.074			19.73	1.73		

Table 5.8: Barra Factor Contributions of the Axioma Portfolios that Produced the Highest Statistically Significant Asset Selection

Source of Return	Average Active Exposure	Contribution (% Return) Average [1]	Variation [2]	Total [1+2]	Total Risk (% Std Dev)	Info Ratio	T-Stat
CURRENCY SENSITIVITY	0.08	0.01	0.02	0.03	0.37	0.08	0.29
EARNINGS VARIATION	0.45	-0.46	0.22	-0.24	0.79	-0.26	-0.89
EARNINGS YIELD	0.09	0.32	0.09	0.41	0.71	0.53	1.85
GROWTH	0.19	-0.13	-0.28	-0.41	0.34	-1.12	-3.88
LEVERAGE	0.44	-0.67	0.19	-0.48	0.74	-0.60	-2.08
MOMENTUM	0.37	-0.76	0.13	-0.63	2.26	-0.30	-1.05
NON-EST UNIVERSE	0.42	0.17	-0.09	0.08	3.06	0.02	0.08
SIZE	-1.17	3.10	0.18	3.29	3.54	0.91	3.16
SIZE NON-LINEARITY	-0.71	-0.55	-0.02	-0.57	2.51	-0.22	-0.75
TRADING ACTIVITY	-0.52	0.10	0.21	0.31	1.06	0.27	0.93
VALUE	0.40	-0.16	-0.23	-0.40	0.83	-0.42	-1.45
VOLATILITY	0.34	0.70	-0.25	0.45	1.80	0.26	0.91
YIELD	0.05	-0.03	-0.12	-0.15	0.19	-0.73	-2.52
Total				1.70	5.72	0.30	1.04

What factor contributions of the Axioma portfolio were rewarded by the market from 1998–2009? Small size (the exposure of -1.17 produced 329 basis points of return with a corresponding t-statistic of 3.16) and earnings yield (the exposure of 0.09 produced 41 basis points of return with a corresponding t-statistic of 1.85, statistically significant at the 10 percent level). Growth, leverage, and yield produced negative statistically significant factor returns in the BARRA USER attributions for 1998–2009.

Why Use the Axioma Statistical Model?

In 2012, MCM ran a second Horse Race competition using a two-I/B/E/S analyst, 7500-largest market-capitalized global universe. The Axioma Statistical Risk Model produced higher geometric means, Sharpe ratios, and information ratios using GLER data than its Fundamental (WW21AxiomaMH) Model as shown in Table 5.9.

Table 5.9: Axioma Analysis of the FactSet Global 7500-stock Data Universe

Axioma Ranked Global Backtest
FSGLER Model using All Country World Growth Index Constituents
Universe: ACWG
Simulation Period: Jan. 1999 - Dec. 2011
Transactions Costs: 150 basis points each w ay, respectively.

Return Model	Risk Model	Tracking Error	Sharpe Ratio	Information Ratio	Ann. Active Return	Ann. Active Risk	N
GLER	STAT	4	0.554	1.475	9.99	6.78	144
		5	0.602	1.385	11.38	8.24	110
		6	0.656	1.409	13.25	9.40	87
		7	0.715	1.454	14.94	10.28	70
		8	0.748	1.451	16.20	11.16	58
	FUND	4	0.382	1.091	6.08	5.57	163
		5	0.460	1.151	7.73	6.72	129
		6	0.521	1.158	9.33	8.06	104
		7	0.582	1.217	11.02	9.06	83
		8	0.647	1.281	12.75	9.95	71

The dominance of the statistical risk model has been a prevailing result of the MCM Horse Races from 2010–2018. Moreover, higher tracking errors produce higher active returns and Sharpe ratios. Investors, over the long run, are compensated for bearing risk. Furthermore, portfolios need to exceed 90–100 stocks unless you are targeting 6–8 percent tracking errors.

5.7 The Axioma Alpha Alignment Factor and Custom Risk Models

Axioma has pioneered two techniques to address the so-called under-estimation of realized tracking errors, particularly during the 2008 Financial Crisis. The first technique, known as the Alpha Alignment Factor (AAF), recognizes the possibility of missing systematic risk factors and makes amends to the greatest extent that is possible without a complete recalibration of the risk model that accounts for the latent systematic risk in alpha factors explicitly. In the process of doing so, the AAF approach not only improves the accuracy of risk prediction, but also makes up for the lack of efficiency in the optimal portfolios.

Custom Risk Model

The second technique, known as the Custom Risk Model (CRM), proposes the creation of a custom risk model by combining the factors used in both the expected-return and risk models, which does not address the factor alignment problem that is due to constraints. Several practitioners have decided to perform a "post-mortem" analysis of mean-variance portfolios, attempted to understand the reasons for the deviation of ex-post performances from ex-ante targets, and used their analysis to suggest enhancements to mean-variance optimization inputs in order to overcome the discrepancy. Lee and Stefek (2008) and Saxena and Stubbs (2012) define this as a factor alignment problem (FAP), which arises as a result of the complex interactions between the factors used for forecasting expected returns, risks and constraints.[24]

While predicting expected returns is exclusively a forward-looking activity, risk prediction focuses on explaining the cross-sectional variability of returns, mostly by using historical data. Expected-return modelers are interested in the first moment of the equity return process, while risk modelers focus on the second moments. These differences in ultimate goals inevitably introduce different factors for expected returns and risks. Even for the "same" factors, expected-return and risk modelers may choose different definitions for good reasons. Constraints play an important role in determining the composition of the optimal portfolio. Most real-life quantitative strategies have other constraints that model desirable characteristic of the optimal portfolio. For example, a client might be reluctant to invest in stocks that benefit from alcohol, tobacco, or gambling activities on ethical grounds, or she might constrain her portfolio turnover so as to reduce her tax burden.

The naïve application of the portfolio optimization has the unintended effect of magnifying the sources of misalignment. The optimized portfolio underestimates the unknown systematic risk of the portion of the expected returns that is not aligned with the risk model. Consequently, it overloads the portion of the expected return that is uncorrelated with the risk factors.

Alpha Alignment Factor

The empirical results in a test-bed of real-life active portfolios based on client data show clearly that the above-mentioned unknown systematic risk is a significant portion of the overall systematic risk, and should be addressed accordingly. Saxena and Stubbs (2012) reported that the earning-to-price (E/P) and book-to-price (B/P) ratios used in the USER Model and Axioma Risk Model have average misalignment coefficients of 72 percent and 68 percent, respectively. While expected-return and risk models are indispensable components of any active strategy, there is also a third component, namely the set of constraints that is used to build a portfolio. Saxena and Stubbs (2012) proposed that the risk variance-covariance matrix C be augmented with additional auxiliary factors in order to complete the risk model. The augmented risk model has the form of

$$C_{new} = C + \sigma_{\underline{\alpha}}^2 \underline{\alpha} \cdot \underline{\alpha}' + \sigma_{\underline{\gamma}}^2 \underline{\gamma} \cdot \underline{\gamma}' \qquad (44)$$

where $\underline{\alpha}$ is the alpha alignment factor (AAF), σ_α is the estimated systematic risk of $\underline{\alpha}$, $\underline{\gamma}$ is the auxiliary factor for constrains, and σ_γ is the estimated systematic risk of $\underline{\gamma}$. The alpha alignment factor $\underline{\alpha}$ is the unitized portion of the uncorrelated expected-return model, i.e., the orthogonal component, with risk model factors. Saxena and Stubbs (2012) reported that the AAF process pushed out the traditional risk model-estimated efficient frontier. Saxena and Stubbs (2015) refer to implied alpha as the unitized portion of the uncorrelated expected-return model in the augmented regression model. Saxena and Stubbs (2015) report that there is a small increment to specific risk compared to its true systematic risk.

Saxena and Stubbs (2012) applied their AAF methodology to the USER model, running a monthly backtest based on the above strategy over the time period 2001–2009 for various tracking error values of σ chosen from {4%, 5%... 8%}. For each value of σ, the backtests were run on two setups, which were identical in all respects except one, namely that only the second setup used the AAF methodology (σ_α = 20%). Axioma's fundamental medium-horizon risk model (US2AxiomaMH) is used to model the active risk constraints.

Saxena and Stubbs (2012) analyzed the time series of misalignment coefficients of alpha, implied alpha, and the optimal portfolio, and found that almost 40–60 percent of the alpha is not aligned with the risk factors. The alignment characteristics of the implied alpha are much better than those of the alpha. Among other things, this implies that the constraints of the above strategy, especially the long-only constraints, play a proactive role in containing the misalignment issue. In addition, not only do the orthogonal components of both the alpha and the implied alpha have systematic risk, but the magnitude of the systematic risk is comparable to that of the systematic risk associated with a median risk factor in US2AxiomMH.

Saxena and Stubbs (2012) showed the predicted and realized active risks for various risk target levels, and noted the significant downward bias in risk prediction when the AAF methodology is not used.[25] The realized risk-return frontier demonstrates that not only does using the AAF methodology improve the accuracy of the risk prediction, it also moves the ex-post frontier upward, thereby giving ex-post performance improvements. In the process of doing so, the AAF approach not only improves the accuracy of risk prediction, but also makes up for the lack of efficiency in the optimal portfolios.[26] Saxena and Stubbs (2015) extended their 2012 *Journal of Investing* research and reported positive frontier spreads.

Sivaramakrishnan and Stubbs (2013) proposed the creation of an Axioma custom risk model by combining the factors used in both the expected-return and risk models, which does not address the factor alignment problem that is due to constraints. The Sivaramakrishnan and Stubbs model allowed great interaction with clients to produce several variations on risk models that were consistent with particular clients' needs of risk exposures. Ceria, Sivaramakrishnan and Stubbs (2015) reported how each alpha signal can be transformed into a factor mimicking portfolio, and how the alpha signals can be combined into a target portfolio with a mean-variance optimization (MVO) problem. The new (combined) alpha signal is constructed as the implied alpha of the target portfolio and used with a custom risk model. The portfolio is consistent as it satisfies the relevant implementation constraints.

Axioma AAF Applications

As we previously noted, Saxena and Stubbs (2012) analyzed the time series of misalignment coefficients of alpha, implied alpha and the optimal portfolio, and found that almost 40–60 percent of the alpha is not aligned with the risk factors. Additionally, not only do the orthogonal components of both the alpha and the implied alpha have systematic risk, but the magnitude of the systematic risk is comparable to that of the systematic risk associated with a median risk factor in US2AxiomMH.

Let us take a deep-dive into the CTEF and GLER variables and their effectiveness. Let us examine the use of AAF in two analyses. First, Guerard (2012) used AAF in creating global portfolios using GLER for the 1999–2009 time period. The Axioma Statistical Risk Model with an AAF of 20 percent produced realized tracking errors closer to targeted tracking errors and higher Information Ratios using GLER data than its Statistical Risk Model without AAF Model. See Table 5.10 for support for Saxena and Stubbs' (2012) position on AAF pushing out the frontiers.

Table 5.10: An AAF Analysis of the Global Expected Returns (GLER) Model

Initial Axioma WRDS GLER Backtest

Universe: ACWG

Simulation Period: Jan. 1999 - March 2009

Transactions Costs: 150 basis points each way, respectively.

Return Model	Risk Model	Tracking Error	Sharpe Ratio	Information Ratio	Ann. Active Return	Ann. Active Risk	N
GLER	STAT	4	0.448	1.247	8.72	6.99	216
		5	0.511	1.119	10.52	8.77	204
		6	0.516	1.089	11.02	10.12	188
		7	0.552	1.074	12.29	11.44	185
		8	0.605	1.111	14.14	12.73	177

AAF

Tracking Error	Sharpe Ratio	Information Ratio	Ann. Active Return	Ann. Active Risk	N
4	0.290	1.159	4.79	4.14	516
5	0.359	1.230	6.37	5.18	442
6	0.397	1.145	7.43	6.49	383
7	0.464	1.179	9.09	7.71	340
8	0.532	1.236	10.94	8.86	304

Second, Guerard and Chettiappan (2017) used AAF in stock selection in Non-US, Global, and EM markets, 2003–2016. We modified the Guerard and Chettiappan (2017) results and report similar statistics in Table 5.11. CTEF and GLER are highly statistically significant in producing statistically significant

Active and Specific returns. Note, however, that EM portfolios dominate the Non-US and Global portfolios in terms of Active Returns and Specific Returns, although all three universes produce statistically significant Active Returns and Specific Returns.

Table 5.11: Forecasted Earnings Acceleration and Stock Selection Modeling in Global, Non-US, and EM Universes

Model: Global CTEF GLER TE Period: 2003-01-31 to 2016-12-30 (Monthly)		Benchmark: ACWG			
	Portfolio	Active		Specific	
	Return	Return	T-Stat	Return	T-Stat
GLOBAL_CTEF-TE4	12.44%	3.89%	2.28	1.36%	1.28
GLOBAL_CTEF-TE6	15.87%	7.32%	3.32	3.89%	2.89
GLOBAL_CTEF-TE8	16.31%	7.77%	3.13	4.13%	2.92
GLOBAL_GLER-TE4	13.28%	4.73%	2.95	1.54%	1.42
GLOBAL_GLER-TE6	16.03%	7.49%	3.61	3.26%	2.44
GLOBAL_GLER-TE8	16.87%	8.32%	3.45	4.27%	2.98

Model: XUS CTEF GLER TE Period: 2003-01-31 to 2016-12-30 (Monthly)		Benchmark: ACWXUSG			
	Portfolio	Active		Specific	
	Return	Return	T-Stat	Return	T-Stat
XUS_CTEF_TE4	14.11%	6.17%	3.59	3.55%	3.10
XUS_CTEF_TE6	16.27%	8.32%	3.89	3.87%	2.85
XUS_CTEF_TE8	18.05%	10.10%	4.28	4.78%	3.15
XUS_GLER_TE4	13.78%	5.83%	3.67	1.93%	1.81
XUS_GLER_TE6	15.11%	7.17%	3.52	3.10%	2.27
XUS_GLER_TE8	16.22%	8.28%	3.82	4.36%	2.94

Model: EM CTEF GLER TE Period: 2003-01-31 to 2016-12-30 (Monthly)		Benchmark: EMG			
	Portfolio	Active		Specific	
	Return	Return	T-Stat	Return	T-Stat
EM_CTEF_TE4	16.15%	5.43%	3.37	4.08%	3.27
EM_CTEF_TE6	17.34%	6.62%	3.00	3.76%	2.27
EM_CTEF_TE8	19.76%	9.04%	3.67	5.17%	2.97
EM_GLER_TE4	17.72%	7.00%	4.90	4.69%	3.97
EM_GLER_TE6	19.81%	9.09%	4.59	6.21%	3.91
EM_GLER_TE8	20.92%	10.20%	4.43	6.55%	3.81

Moreover, AAF helps enhance IRs at targeted tracking errors of 5, 6, and 7 percent, respectively, with the Axioma Fundamental Risk Model in the EM universe and targeted tracking errors of 5 and 7 percent with the Axioma Statistical Risk Model. See Table 5.12.

Table 5.12: A Deep Dive into EM Portfolio Construction, 2003–2016

	FUND Risk Model			
	Tracking Error			
Model: EM MQ	5.00	6.00	7.00	10.00
Ann. Port Return	16.76	17.34	18.36	20.62
Ann STD	22.06	22.19	22.26	22.40
Ann. Active Return	6.45	7.24	8.05	10.31
Ann. Active Risk	5.04	5.95	6.65	8.35
N	68.20	63.40	58.50	55.40
ShR	0.682	0.714	0.748	0.844
IR	1.281	1.217	1.208	1.235
	FUND AAF20% Risk Model			
	Tracking Error			
Model: EM MQ	5.00	6.00	7.00	10.00
Ann. Port Return	16.53	17.19	18.87	10.44
Ann STD	22.11	22.09	22.22	22.35
Ann. Active Return	6.29	6.89	8.57	10.12
Ann. Active Risk	4.31	5.24	6.94	8.15
N	77.80	69.70	59.80	56.00
ShR	0.671	0.701	0.773	0.838
IR	1.447	1.309	1.234	1.242
	FUND AAF40% Risk Model			
	Tracking Error			
Model: EM MQ	5.00	6.00	7.00	10.00
Ann. Port Return	DNC	DNC	DNC	19.51
Ann STD	DNC	DNC	DNC	22.26
Ann. Active Return	DNC	DNC	DNC	9.21
Ann. Active Risk	DNC	DNC	DNC	7.17
N	DNC	DNC	DNC	61.90
ShR	DNC	DNC	DNC	0.800
IR	DNC	DNC	DNC	1.284
	STAT Risk Model			
	Tracking Error			
Model: EM MQ	5.00	6.00	7.00	10.00
Ann. Port Return	17.38	17.80	18.55	19.79
Ann STD	22.57	22.44	22.42	22.73
Ann. Active Return	7.08	7.50	8.25	9.48
Ann. Active Risk	5.04	5.95	6.65	8.35
N	64.30	60.90	56.60	53.50
ShR	0.695	0.717	0.751	0.796
IR	1.271	1.146	1.103	1.075
	STAT AAF20% Risk Model			
	Tracking Error			
Model: EM MQ	5.00	6.00	7.00	10.00
Ann. Port Return	16.93	DNC	17.84	19.39
Ann STD	22.56	DNC	22.48	22.74
Ann. Active Return	6.62	DNC	7.54	9.08
Ann. Active Risk	4.76	DNC	62.10	8.54
N	75.90	DNC	62.10	56.90
ShR	0.675	DNC	0.717	0.778
IR	1.391	DNC	1.131	1.064
DNC = Did Not Converge				

The application of the Axioma Fundamental Risk Model to the EM universe produced a real-time portfolio having Geometric Means, Sharpe Ratios, and IRs in the top decile of EM portfolios, see Guerard and Chettiappan (2017). See Table 5.13.

Table 5.13: EM Real-Time Attribution Analysis

Portfolio: McKinley Capital Emerging Markets Growth	**Base Currency:** USD
Benchmark: MSCI EM (EMERGING MARKETS)	**Return Scaling:** Annualized (Geometric)
Period: 2011-03-31 to 2017-09-29 (Monthly)	**Risk Type:** Realized Risk
Risk Model: WW21AxiomaMH	**Long/Short:** Long Only

Source of Return	Contribution	Avg Exposure	Hit Rate	Risk	IR	T-Stat
Portfolio	**8.18%**			**16.11%**		
Benchmark	**1.94%**			**17.26%**		
Active	**6.24%**	**0.00%**		**5.35%**	**1.17**	**2.97**
Specific Return	**2.77%**	**0.00%**		**3.36%**	**0.82**	**2.10**
Factor Contribution	**3.47%**	**0.00%**		**4.35%**	**0.80**	**2.03**
Style	**1.24%**	**0.9584**		**2.76%**	**0.45**	**1.14**
Exchange Rate Sensitivity	-0.07%	0.0278	38.46%	0.07%	-1.05	-2.68
Growth	0.32%	0.3058	67.95%	0.33%	0.96	2.45
Leverage	0.05%	-0.0788	55.13%	0.12%	0.44	1.13
Liquidity	-0.16%	0.1390	50.00%	0.22%	-0.71	-1.82
Medium-Term Momentum	4.03%	0.7433	74.36%	2.23%	1.80	4.60
Short-Term Momentum	-0.78%	0.1210	21.79%	0.41%	-1.91	-4.86
Size	0.05%	-0.1451	52.56%	0.50%	0.09	0.24
Value	-0.71%	-0.2992	42.31%	0.56%	-1.27	-3.23
Volatility	-1.49%	0.1447	34.62%	0.76%	-1.94	-4.96
Country	**0.59%**	**-4.21%**		**1.04%**	**0.57**	**1.45**
Industry	**1.76%**	**-4.21%**		**1.17%**	**1.50**	**3.82**
Currency	**0.10%**	**-1.25%**		**0.81%**	**0.12**	**0.31**
Local	**0.04%**	**0.36%**		**0.14%**	**0.31**	**0.78**
Market	**-0.25%**	**-4.21%**		**0.66%**	**-0.39**	**-0.99**
Sectors	**1.76%**	**-4.21%**		**1.17%**	**1.50**	**3.82**

The portfolio techniques outperform when implemented with lambdas of 200 or targeted tracking errors of 5–7 percent.[27]

A final word of caution could be appropriate. We believe that quantitative portfolio construction should outperform the universe benchmark over 5- to 30-year investment horizons as long as models are applied consistently. Andrew Lo (2017), a noted and delightful professor of financial economics at MIT, has noted in his *Adaptive Markets* that a new investment paradigm might occur when investors are dominated by individuals facing extreme financial threats and that risk might be unduly punished, offsetting the tradition risk/return tradeoff. Moreover, changing economic dynamics might be more important that factor models in assessing risk, particularly in the post-2000 time period.

5.8 Assessing Mutual Funds: The Treynor Index

This section draws heavily from Lee, Lee, and Liu (2010) and reviews the selectivity, market timing, and overall performance of equity funds in the United States during January 1990 until September 2005.[28] Lee, Lee, and Liu used Sharpe, Treynor, and Jensen measures to evaluate the selectivity performance of mutual fund managers. In addition, we also used the Treynor-Mazuy and Lee-Rahman models to evaluate the selectivity and timing performance of mutual fund managers. Based on these measures and models, Lee, Lee, and Liu reported that about one-third of funds had significantly positive selectivity ability, and some had timing ability for the mutual fund managers. Nevertheless, without considering transaction costs and taxes, the actual investment for most mutual funds compared to a passive investment strategy still appears to take the lead.

The investment of mutual funds has been extensively studied in finance. Over the last few decades, there has been a dramatic increase in the development of instruments measuring the performance of mutual funds. Early researchers (Treynor (1965), Sharpe (1966), and Jensen (1968)) used a one-parameter indicator to evaluate the portfolio performance. However, these studies assume the risk levels of the examined portfolios to be stationary through time. Fama (1972) and Jensen (1972) pointed out that the portfolio managers can adjust their risk composition according to their anticipation for the market. Moreover, Fama (1972) suggested that the managers' forecasting skills can be divided into two parts: the selectivity ability and the market timing ability. The former is also named as micro-forecasting, involving the identification of the stocks that are under- or over-valued relative to the general stocks. The latter is also named as macro-forecasting, involving the forecast of future market return. In other words, the selectivity and market timing abilities of fund managers are viewed as important factors deciding the overall fund performance.[29]

Treynor and Mazuy (1966) used a quadratic term of the excess market return to test for market timing ability. It can be viewed as the extension of the Capital Asset Pricing model (CAPM). If the fund manager can forecast market trend, he will change the proportion of the market portfolio in advance. Jensen (1972) developed the theoretical structure for the timing ability. Under the assumption of a joint normal distribution of the forecasted and realized returns, Jensen showed that the correlation between the managers' forecast and the realized return can be used to measure the timing ability. In addition, Henriksson and Merton (1981) used options theory, developed by Merton (1981), to explain the timing ability.

Models

Lee, Lee, and Liu (2010) empirically examined the mutual fund performance by using six models proposed respectively by Treynor (1965), Sharpe (1966), Jensen (1968), Treynor and Mazuy (1966), Henriksson and Merton (1981), and Lee and Rahman (1990). The monthly returns for 189 months (January 1990 to September 2005) for a sample of 628 open-end equity funds were used.

Treynor Index

Treynor (1965) uses the concept of the security market line[30] drawn from the CAPM to get a coefficient β. Under the assumption of complete diversification of asset allocation, it means that we just have systematic risk measured by β. The Treynor index (TI) measuring the reward per unit of systematic risk for the portfolio can be shown as follows:

$$TI = \frac{\bar{r}_p - R_f}{\beta_p} \tag{45}$$

where $\bar{r}_p$ is the average return of the pth mutual fund, and r_f is defined as risk-free rate. The numerator of Treynor index can be viewed as excess return on the portfolio. This ratio is a risk-adjusted performance value. This indicator is suitable for valuing the performance of a well-diversified portfolio; this is because it just takes the systematic risk into account.

Sharpe Index

Different from Treynor (1965), Sharpe (1966) argues the phenomenon that the fund managers will be in favor of fewer stocks. Therefore, it is impossible to diversify the individual risks completely. In other words, the excess return should be calculated based on the total risk (including systematic and nonsystematic risks). The Sharpe Index (SI), applying the concept of the capital market line[31] can be written as:

$$SI = \frac{\bar{\mu}_p - R_f}{\sigma_p} \tag{46}$$

where σ_p is the standard deviation of the portfolio, namely total risk. The Sharpe index is expressed as the reward per unit of total risk. The higher the two indices mentioned above, the better the fund's performance. Because this measure is based on the total risk, it enables measurement of the performance of the portfolio, which is not very diversified.

Jensen Index

Jensen (1968) proposes a regression-based view to measure the performance of the portfolio. The Jensen index (or called Jensen alpha) uses the CAPM to determine whether a fund manager outperformed the market. Its formula is as follows:

$$R_{p,t} = \alpha_p + \beta_{p,t} R_{m,t} + u_{p,t} \tag{47}$$

where $R_{p,t}$ and $R_{m,t}$ are the excess returns ($R_t = r_t - r_f$) at time t of the portfolio return and the market return, respectively. The term of $u_{p,t}$ in the formula is the residual at time t. The coefficient α_p is used to measure the performance of mutual funds in the sense of the additional return due to the manager's choice. It also represents the fund manager's selectivity ability without considering timing ability. A significantly positive and high value of Jensen alpha indicates superior performance compared with the market index.

Treynor-Mazuy Model

Treynor and Mazuy (1966), putting a quadratic term of the excess market return into equation (28), provide us with a better framework for the adjustments of the portfolio's beta to test a fund manager's timing ability. The fund manager with timing ability will be able to adjust the risk exposure from the market. To take a simple example, if a fund manager expects a coming up (down) market, he will hold a larger (smaller) proportion of the market portfolio. Therefore, the portfolio return can be viewed as a convex function of the market return. The equation can be given below:

$$R_{p,t} = \alpha_p + \beta_1 R_{m,t} + \beta_2 R_{m,t}^2 + \varepsilon_{p,t} \tag{48}$$

where the coefficient β_2 is used to measure the timing ability. When β_2 is significantly larger than zero, it represents that, in an up (down) market, the increasing (decreasing) proportion in the risk premium of the mutual fund is larger than that in the market portfolio. This model was formulated empirically by Treynor and Mazuy (1966) and validated by Jensen (1972).

Merton-Henriksson Market Timing Model

Henriksson and Merton (1981) used options theory to explain the timing ability. It consists of a modified version of the CAPM that takes the manager's two objectives into account and depends on whether he forecasts that the market return will or will not be better than the risk-free asset return. They view the coefficient β as a binary variable. This means that a fund manager with market timing ability should have different β values in the up and down markets.[32] We can express the equation as:

$$R_{p,t} = \alpha_p + \beta_1 R_{m,t} + \beta_2 Max(0, -R_{m,t}) + \varepsilon_{p,t} \tag{49}$$

If $\beta_2 > 0$, this shows that the manager is able to forecast a market to be down or up. For an up market (a down market), the equation (49) can be expressed as

$$R_{p,t} = \alpha_p + \beta_1 R_{m,t} + \varepsilon_{p,t} \; (R_{p,t} = \alpha_p + (\beta_1 - \beta_2) R_{m,t} + \varepsilon_{p,t}) \tag{50}$$

Jensen (1972) showed that the timing ability can be measured by the correlation between the managers' forecast and the realized return. Pfleiderer and Bhattacharya (1983) modify Jensen's (1972) model[33] to propose a regression-based model to evaluate the market timing and selectivity abilities.[34]

Assessment of Funds

Lee, Lee, and Liu (2010) used the alternative methods to examine the selectivity, market timing, and overall performance for the open-end equity mutual funds.[35] The samples used were the monthly returns of the 628 mutual funds[36] ranging from January 1990 to September 2005 with 189 monthly observations. The fund data were obtained from the CRSP Survivor-Bias-Free US Mutual Fund Database. Then, they used the ICDI's fund objective codes to sort the objectives of the mutual funds. In total, there are 23 types of mutual fund objectives; simplifying into as two groups, growth funds and non-growth funds,[37] consisting of 439 growth funds and 189 non-growth funds. In addition to the CRSP fund data, the S&P 500 stock index obtained from Datastream is used for the return of the market portfolio. Moreover, Lee, Lee, and Liu use the Treasury bill rate with a 3-month holding period as the risk-free return. The Treasury bill rate is available from the website of the Federal Reserve Board.

Lee, Lee, and Liu (2010) reported that the difference of monthly returns of growth and non-growth funds was quite large. The result for the 1990–2005 period shows that 82 percent (69 percent) of the growth (non-growth) funds have better performance than the market. This seems to point out the growth funds are more valuable to be invested than the non-growth funds.[38] Kosowski, Timmermann, Wermers, and White (2006) reported that the top 5 percent of abnormal-growth funds produced persistent and statistically significant excess returns relative to the four-factor Fama-French (1992) and Carhart (1997) risk factors for 2118 open-end US equity funds for the January 1975–December 2002 period. Persistence was determined by two periods, the 1975–1989 and 1990–2002 periods. Moreover, Kosowski, Timmermann, Wermers, and White (2006) reported that the worst two deciles of abnormal-growth significantly underperformed the universe of funds. They found no evidence of "stars" or significant and persistent outperformance in income funds.

The performance of the Treynor index is mainly based on the systematic risk obtained from the CAPM. Recent work by Berk and Von Binsbergen (2016 and 2017) has reported the dominance of the Treynor index effectively assessing mutual fund performance. It is appropriate to evaluate the well-diversified portfolio. Different from the Treynor index, the Sharpe index takes account of the total risk of the portfolio. Lee, Lee, and Liu (2010) reported that 30 percent of the growth funds for the entire period have Jensen alpha significantly larger than zero with 95 percent confidence.

The first three indicators assume that the portfolio risk is stationary and only take the stock selection into account. If we want to modify the level of the portfolio's exposure to the market risk, the market timing ability must be demonstrated. The models with the ability to test market timing are against the CAPM. This is due to permit variations in the portfolio's beta over the investment period. There are three models for evaluating the selectivity and timing abilities in this study. The Treynor-Mazuy model, a quadratic version of the CAPM, tests market timing abilities. Lee, Lee, and Liu reported that over 80 percent of the funds have negative values of timing ability and a considerable ratio among them have significantly negative estimates. For the selectivity ability, a very high ratio of the funds has positive estimates and many of them are significantly positive. Only very few funds have significantly negative estimates. The Henriksson-Merton model is also a modified version of the CAPM that applies a binary choice for the manager. It depends on whether the market return will or will not perform better than the risk-free asset return. The estimates of the selectivity ability even display larger values. Furthermore, the correlations between the estimates of timing and selectivity ability are -0.85 for the entire period. The Lee-Rahman model assumes no negative timing ability. The non-growth funds have better timing ability than the growth funds for all periods. Different from the previous two models, the correlations between the estimates of timing and selectivity ability are 0.43 for the entire period. Moreover, 40 of the funds in the entire period have significantly positive estimates in both selectivity and timing ability.

What do we know about mutual funds and their relative performance? Bogle (2000, 2009) tells us that actively managed mutual funds underperform the market by approximately 200 basis points. Lee, Lee, and Liu (2010) reported that growth funds outperformed the market for the 1995–2005 period. Kosowski, Timmermann, Wermers, and White (2006) and Wei, Wermers, and Yao (2016) reported that contrarian funds outperformed the market during the 1995–2012 period, and Berk and van Binsbergen (2016a, 2016b) demonstrate that the Treynor CAPM model dominates the Fama-French (1992, 1995) and Carhart (1997) models in assessing mutual fund performance for the 1995–2011 period.

5.9 What Have You Done for Me Lately?

Do we, the authors, believe in efficient markets? No, of course, not. We have shown that statistically significant active returns can be earned over 5- and 10-year periods. Mean-Variance tracking error at risk portfolios using CTEF, forecasted earnings acceleration, GLER, and corporate exports variables work in backtest and real-time performance. Markowitz has always asked researchers, "What have you done for me lately?" Guerard and Markowitz (2018) address that question. Using the Quant models presented in this chapter, Guerard and Markowitz reported that CTEF, GLER, and CE have outperformed, on average, and in 70–77 percent of the years post-2003, using ITG transactions costs curve, and post-publication of the models. Using a large universe of global stocks, the authors cross validated their previous results. The only requirements are intelligence and patience. You will not win in all years. You need about five years to determine statistical significance. What do you do if your mutual fund has a Sharpe Ratio less than the stock index (S&P 500 or Russell 3000), negative Information Ratios, and Treynor Ratios? Sell it. Insignificant active returns and negative specific returns should never be tolerated. In life there are winners and losers. In investments, winners produce statistically significant active returns and positive specific returns over 5-year, 10-year, and since-inception periods.

5.10 Summary and Conclusions

This chapter addresses several aspects of risk, return, and performance measurement. It is important to see how quantitative analysis was developed in investment research and analysis. Harry Markowitz, Bill Sharpe, and Jack Treynor pioneered capital market equilibrium and the creation and estimation of the Capital Asset Pricing Model. In the 1970s, Barr Rosenberg and his colleagues at Barra developed and estimated multi-factor risk models. Recent research and commercialization by APT and Axioma has furthered portfolio optimization. The APT Tracking Error at Risk and the Axioma Alpha Alignment Factor Models have pushed out the Efficient Frontiers. Mutual fund performance measurement, pioneered by Sharpe and Treynor, incorporates risk estimation. Recent mutual fund studies by Kosowski, Timmermann, Wermers, and White, Wei, Wermers, and Yao, and Berk and van Binsbergen demonstrate that excess returns can be earned. Do the authors believe in efficient markets? No, of course, not. We have shown that statistically significant active returns can be earned over 5 and 10-year periods. Markowitz has always asked researchers, "What have you done for me lately?" Guerard and Markowitz (2018) answer the question. We have shown the readers variables to enhance returns over the benchmark by 100-plus basis points. That was, and is, the stated goal of this text.

---

[1] The authors appreciate comments of Professors C.F. Lee and Bernell Stone, Dr. Anureet Saxena, and Dr. Jose Menchero on earlier drafts of this chapter. Any errors remaining are the sole responsibility of the primary author.

[2] Jack Treynor (1962, 1999), following in the steps of Markowitz and Modigliani and Miller (1958), sought to provide the groundwork for a theory of valuation that incorporates risk in his "Toward a Theory of Market Value of Risky Assets." Treynor considered an idealized capital market to establish how risk premiums implicit in share prices are related to portfolio decisions of investors without the complexities of taxes and other frictions that can significantly affect share prices in the real world. Treynor listed his assumptions: (1) no taxes; (2) no frictions, such as brokerage costs; (3) the individual investor cannot influence the price; (4) investors maximize expected utility, with primary concern for the first and second moments of the distribution of outcomes; (5) investors are risk-averse; (6) a perfect lending market exists; and (7) perfect knowledge, interpreted to mean knowledge of present prices, and what investors know about the future might have a bearing of future investment values. Treynor stated that the expected yield to the investor is a return on his capital at the risk-free lending rate and an expected return for risk-taking. See Treynor (2008, pp. 49–51).

[3] Treynor, 2008, p. 57.

[4] *The Founders of Modern Finance: Their Prize-Winning Concepts and 1990 Nobel Lectures*, The Research Foundation of the Institute of Chartered Financial Analysts. 1991.

[5] Bill Sharpe (fn. 7, pp. 77) noted that upon completion of his paper, he learned that Mr. Jack L. Treynor, of Arthur D. Little, Inc., had independently developed a model similar in many respects to his. Jack's excellent work, he noted is at present unpublished. In summary, Treynor's portfolio returns were proportional to the covariances among the stocks and market index, not the total risk of the stocks. The covariances among stocks and the market index would be recognized in the coming months as stock betas by Bill Sharpe.

[6] According to BARRA online advertisements.

[7] There are several definitive treatments of the Barra system. Rosenberg and Marathe (1979) was a seminal test of capital asset pricing. Rudd and Clasing (1982), Grinold and Kuhn (2000), and Menchero, Morozov, and Shepard (2010) are some of the most cited Barra model publications.

[8] Barr Rosenberg and several co-authors put forth some 12–15 papers during the 1973–1979 time period that created the intellectual basis of the Barra system. The Rosenberg publications were both academic and practitioner journals. The Rosenberg and Marathe (1979) paper, in the mind of the authors, represents the definitive multi-factor risk model analysis / test of the 1970s.

[9] Markowitz discusses the MFM formulation in his second monograph, *Mean-Variance Analysis in Portfolio Choice and Capital Markets*, New Hope, PA: Frank J. Fabozzi Associates, 2000, Chapter 3, pp. 45–47. The Markowitz (1987, 2000) Mean-Variance Analysis volume requires great patience and thought on the part of the reader, as noted by Bill Sharpe in his foreword to the 2000 edition.

[10] Jose Menchero and his colleagues at BARRA authored the "Global Equity Risk Modeling" article in the Markowitz volume, estimated an eight-risk global index model in the spirit of the Rosenberg USE3 model, see Guerard (2010).

[11] We have glossed over a number of econometric subtleties in these few sentences. Those readers who wish to learn more about these estimation difficulties are directed toward the following articles and the references contained there: Merton Miller and Myron Scholes, "Rates of Return in Relation to Risk: A Reexamination of Recent Findings," in *Studies in The Theory of Capital Markets*, ed. Michael Jensen (New York: Praeger Publishers, 1972), pp. 47–48, and Eugene F. Fama, *Foundations of Finance* (New York: Basic Books, 1976), Chapter 4.

[12] When an analyst forms a judgment on the likely performance of a company, many sources of information can be synthesized. For instance, an indication of future risk can be found in the balance sheet and the income statement; an idea as to the growth of the company can be found from trends in variables measuring the company's position; the normal business risk of the company can be determined by the historical variability of the income statement; and so on. The approach that Rosenberg and Marathe take is conceptually similar to such an analysis since they attempt to include all sources of relevant information. This set of data includes historical technical and fundamental accounting data. The resulting information is then used to produce, by regression methods, the fundamental predictions of beta, specific risk, and the exposure to the common factor (dummy) variables for 39 industry groups as the method of introducing industry effects.

[13] The COMPUSTAT database is one of the databases collected by Investors Management Sciences, Inc., a subsidiary of Standard & Poor's Corporation.

[14] See Barr Rosenberg and Vinay Marathe, "Common Factors in Security Returns: Microeconomic Determinants and Macroeconomic Correlates," *Proceedings of the Seminar on the Analysis of Security Prices*, University of Chicago, May 1976, pp. 61-115.

[15] The result from the cross-sectional regression equation (27) is the specific return and specific risk on the security, together with the 45 coefficients. These estimated coefficients represent the returns that can be attributed to the factors in the month of the analysis.

[16] Rudd and Clasing (1982) note that the regression coefficient on the market and the regression coefficient on the factor (i.e., b_j and β_j, and b_P and β_P) are close but not identical. In other words, for well-diversified portfolios, the majority of institutional portfolios, one can approximate the portfolio beta by its regression coefficient on the factor, and vice versa, that is, $\beta_P \cong b_P$. In a multiple factor model, the security beta is a weighted average of the factor betas and the beta of the specific return of the security, where the weights are simply the factor loadings for the jth security.

[17] The USE3 extra-market factor was composed of:

1. **Volatility** is composed of variables including the historic beta, the daily standard deviation, the logarithm of the stock price, the range of the stock return relative to the risk-free rate, the options pricing model standard deviation, and the serial dependence of market model residuals.
2. **Momentum** is composed of a cumulative twelve-month relative strength variable and the historic alpha from the 60-month regression of the security excess return on the S&P 500 excess return.
3. **Size** is the log of the security market capitalization.
4. **Size Nonlinearity** is the cube of the log of the security market capitalization.
5. **Trading Activity** is composed of annualized share turnover of the past five years, twelve months, quarter, and month, and the ratio of share turnover to security residual variance.
6. **Growth** is composed of the growth in total assets, five-year growth in earnings per share, recent earnings growth, dividend payout ratio, change in financial leverage, and analyst-predicted earnings growth.
7. **Earnings Yield** is composed of consensus analyst-predicted earnings to price and the historic earnings to price ratios.
8. **Value** is measured by the book to price ratio.

9. **Earnings Variability** is composed of the coefficient of variation in five-year earnings, the variability of cash flow, and the variability of analysts' forecasts of earnings to price.
10. **Leverage** is composed of market and book value leverage, and the senior debt ranking.
11. **Currency Sensitivity** is composed of the relationship between the excess return on the stock and the excess return on the S&P 500 Index. These regression residual returns are regressed against the contemporaneous and lagged returns on a basket of foreign currencies.
12. **Dividend Yield** is the Barra-predicted dividend yield.
13. **Non-Estimation Universe Indicator** is a dummy variable which is set equal to zero if the company is in the Barra estimation universe and equal to one if the company is outside the Barra estimation universe.

[18] Jose Menchero, D.J. Orr, and Jun Wang, "The Barra US Equity Model (USE4): Methodology Notes," August 2011. The main advances of USE4 are:

1. An innovative eigenvector risk adjustment that improves risk forecasts for optimized portfolios by reducing the effects of sampling error on the factor covariance matrix
2. A Volatility Regime Adjustment designed to calibrate factor volatilities and specific risk forecasts to current market levels
3. The introduction of a country factor to separate the pure industry effect from the overall market and provide timelier correlation forecasts
4. A new specific risk model based on daily asset-level specific returns
5. A Bayesian adjustment technique to reduce specific risk biases due to sampling error
6. A uniform responsiveness for factor and specific components, providing greater stability in sources of portfolio risk
7. A set of multiple industry exposures based on GICS®
8. An independent validation of production code through a double-blind development process to assure consistency and fidelity between research code and production code
9. A daily update for all components of the model

[19] The Barra Global Equity Model, GEM2, offered statistically significant results for optimized Value, Momentum, Liquidity, and Size risk factor portfolios. One needed to have a unit exposure to the particular factor, and zero exposures to all other factors. See Jose Menchero, Andrei Morozov, and John Guerard, "Capturing Equity Risk Premia." in C.F. Lee, J. Finnerty. J. Lee, A.C. Lee, and D. Wort, *Security Analysis, Portfolio Management, and Financial Derivatives* (Singapore: World Scientific, 2013), Chapter 25.

[20] See Markowitz (1959), Chapter 9.

[21] Guerard, Krauklis, and Kumar (2012) reported that mean-variance portfolios using the Tracking Error at Risk optimization technique outperform the mean-variance optimization technique during the 1997–2009 period. Both optimization techniques produce statistically significant asset selection. Wormald and van der Merwe (2012) implemented three risk control strategies are implemented. The three strategies are very similar, except for differences in systematic active risk constraints. The first strategy constructs portfolios without any constraints on systematic tracking error (TE), and is referred to as *NoRiskConst.* Another strategy places a mild constraint on systematic TE and is referred to as *MildRiskConst.* The mild constraint level reflects a level of systematic TE slightly lower than the average of the observed values in *NoRiskConst.* In *MildRiskConst* systematic TE is constrained to be below 2.3 percent. The third strategy constrains systematic TE to be below 1.5 percent and is called *StrongRiskConst.* We estimate the systematic risk optimization technique of Systematic Tracking Error optimization technique reported in Wormald and van der Merwe (2012). The APT measure of portfolio risk, TaR, estimates the magnitude that the portfolio return may deviate from the benchmark return over one year. Specifically, the TaR optimization technique emphasizes systematic risk, rather than total risk, in portfolio optimization.

[22] Any optimization system using less than an EAW2 is little no more than a benchmark-hugger, and the passing of the asset manager, unless the manager has assets under management (AUM) of at least $300 billion (USD) will not be mourned as the authors stated in Chapter 1. EAW2 is useful if the asset manager uses a lambda of 1000, see Table 5.4; else, a lambda less than 200 is sub-optimal.

[23] *Axioma Robust Risk Model Handbook*, January 2010.

[24] The author expresses great appreciation for many conversations with Anureet Saxena on this topic.

[25] The bias statistic shown is a statistical metric that is used to measure the accuracy of risk prediction. If the ex-ante risk prediction is unbiased, then the bias statistic should be close to 1.0. Clearly, the bias statistics obtained without the aid of the AAF methodology are significantly above the 95 percent confidence interval, which shows that the downward bias in the risk prediction of optimized portfolios is

statistically significant. The AAF methodology recognizes the possibility of inadequate systematic risk estimation, and guides the optimizer to avoid taking excessive unintended bets.

[26] Guerard, Markowitz, and Xu (2013 and 2015) created efficient frontiers using both of the AxiomaRisk Models, and found that the statistically based Axioma Risk Model, the authors denoted as "STAT", produced higher geometric means, Sharpe ratios, and information ratios than the Axioma fundamental Risk Model, denoted as "FUND". The AAF technique was particularly useful with composite models of stock selection using fundamental data, momentum, and earnings expectations data. Furthermore, the geometric means and Sharpe ratios increase with the targeted tracking errors. However, the information ratios are higher in the lower tracking error range of 3–6 percent, with at least 200 stocks, on average, in the optimal portfolios. The Guerard et al. studies assumed 150 basis points, each way, of transactions costs. The use of ITG cost curves produced about 115–125 basis points of transactions costs, well under the assumed costs. The Guerard et al. studies also used the Sungard APT statistical model, which produced statistically significant asset selection in US and global portfolios.

[27] The Non-US MCM portfolio produces statistically significant Active Returns with positive Specific Returns, about 100 basis points over the corresponding time period, but we do not report the table as we do not want to appear too commercially oriented; we are, we just do not want to appear to be.

[28] C. F. Lee, A. Lee, and N. Liu, "Alternative Model to Evaluate Selectivity and Timing Performance of Mutual Fund Managers: Theory and Evidence", in J. Guerard, Jr. *Handbook of Portfolio Construction: Contemporary Applications of Markowitz Techniques* (New York: Springer, 2010).

[29] However, Brinson, Singer, and Beebower (1991) found that selectivity and market timing abilities only have small influence on fund performance (< 10%). The overall performance should be mostly decided by asset allocation between stock and bond markets.

[30] At equilibrium, all assets are located on this line.

[31] In the presence of a risky asset, this straight line is the efficient frontier for all investors.

[32] Ferson and Schadt (1995) assumed that market prices of securities reflected public information and allowed betas of stocks and portfolios to change with economic conditions. See Christopherson, Carino, and Ferson (2009, Ch. 12) for a more complete discussion of conditional performance evaluation.

[33] In a framework of Jensen (1972), the coefficients in the model cannot be estimated efficiently. However, with some assumptions proposed by Pfleiderer and Bhattacharya (1983), we can get the efficient estimators. The detail can be found in Lee and Rahman (1990), pp. 265-266.

[34] Lee and Rahman (1990) find that in the residual terms of the Pfleiderer and Bhattacharya exists heteroscedasticity. Thus, the estimated coefficients are not efficient. The way to solve this problem is to calculate the variance of the residuals, ω_t and τ_t. Lee and Rahman used a GLS method with correction for heteroscedasticity to adjust the weights of the variables in equation (8) and (10) by σ_ω^2 and σ_τ^2.

[35] In addition to equity funds (the code in CRSP is EQ), Standard & Poor's Main Category provides the other 4 kinds of funds to define fund styles, i.e., fixed income (FI), money market (DG), asset allocation (AA), and convertible (CT), but they are not analyzed in this study.

[36] We delete the funds with any missing values in this period. In addition, the funds with over 20 zero returns are also dropped. This is because we view too many zero values as missing data or lower liquidity for the fund. The list of the fund names is available from the authors on request.

[37] Three types of mutual funds belong to the growth group, i.e. aggressive growth (AG), growth and income (GI), and long-term growth (LG); the other 20 types are put in the non-growth group.

[38] Lee, Lee, and Liu did not consider the transaction costs and taxes here. In general, the growth fund will ask for a higher commission than the non-growth fund.

Chapter 6: Data Mining Corrections

6.1 Introduction to Data Mining

Scientific research is usually a multi-stage process involving:

1. collecting data
2. analyzing collected data
3. finding truth or patterns

Traditional statisticians perform scientific research differently. They will propose the truth first (prior-hypothesis), then collect data and analyze it. Statisticians either reject the prior-hypothesis or shift the hypothesis through Bayesian methods.

Advances in technology make it easier to collect and store data of much larger amounts and of more complexity. The ever-increasing computational power enables human researchers and machines to process this data to find patterns with or without prior hypothesis because they can update the hypothesis so quickly. Those found patterns can be so complex that they would never have been proposed as prior-hypothesis by traditional statisticians.

The process of finding truth without prior-hypothesis got dubious names like "data mining" or "data snooping" in the earlier days. See Lo and MacKinlay (1990) on the subtle data-mining bias in testing the capital asset pricing model covered in Chapter 3. With newly developed statistical methods such as cross validation to verify the "truth" found by "data mining," data mining has been the hottest and fastest growing research area and has impacted every aspect of our lives. One needs only read Lo (2017) to gain tremendous insight into the evolution of financial thought.

Quantitative asset management has been a traditional area for data mining. In 1964, the Center for Research in Security Prices (CSRP) released its database, which contained information about the capital changes and dividends paid for all the stocks listed in the New York Stock Exchange (NYSE) from December of 1925 to the release date. This database makes it possible to calculate daily and monthly total returns for all stocks. Standard & Poor's released its fundamental (the company's financial measures covered in Chapter 2) database Compustat at 1962, which covered 99.9 percent of companies in the world with annual data available back to 1950, and quarterly data available back to 1962.
With such extensive commercial databases available for public companies' fundamental variables and stock returns, it is much easier to do data mining in the asset management area. In Chapter 4, we have shown how to build expected return models based on expected earnings data, valuation ratios, and past stock price performance. We have shown how to construct and manage portfolios by a portfolio optimizer that combines the expected return and risk models in Chapter 5. We have shown that excess return, the Treynor index, Sharpe ratio, and Jensen alpha are appropriate measures of portfolio performance.

The unit tilt financial models used in this book are from three sources and are grouped into four sections in Table 6.1. The first source is the financial report as discussed in Chapter 2. The unit tilt models derived from this source are grouped into the Reported Fundamental Variables section. The second source is forecasted earnings and related information. The unit tilt models derived from this source are grouped into

Forecasted Earnings section. The third source is stock returns. The unit tilt models derived this source are grouped into the Derived from Returns section. This book focuses on how to combine these unit tilt models to create superior composite models, and resulting models are grouped into the Composite Models section.

Table 6.1: Description of Investment Models

Source	Acronym	Description
Reported Fundamental Variables	BP	Book value per share/price per share
	CP	Cash flow per share/price per share
	CSI	Common stock issued
	CSR	Common stock repurchased
	DI	Debt issued
	DP	Dividends Paid/price per share
	DR	Debt repurchased
	EP	Earnings per share/price per share
	ES	Corporation exports
	NDR	Net debt repurchased
	RBP	current BP ratio /average BP ratio over the past five years
	RCP	current CP ratio /average CP ratio over the past five years
	REP	current EP ratio /average EP ratio over the past five years
	ROA1	One-year return on asset
	ROA3	Three-year return on asset
	ROA5	Five-year return on asset
	ROE1	One-year return on equity
	ROE3	Three-year return on equity
	ROE5	Five-year return on equity
	ROIC	Return on invested capital
	SP	Net sales per share/price per share
	RSP	current SP ratio /average SP ratio over the past five years
Forecasted Earnings	BR1	One-year-ahead forecasted earnings per share monthly breath/price per share
	BR2	Two-year-ahead forecasted earnings per share monthly breath/price per share
	FEP1	One-year-ahead forecasted earnings per share/price per share
	FEP2	Two-year-ahead forecasted earnings per share/price per share
	RV1	One-year-ahead forecasted earnings per share monthly revision
	RV2	Two-year-ahead forecasted earnings per share monthly revision
Derived from Returns	Alpha	Regression alpha of excess return to market excess return
	PM	$Price_{t-1}/Price_{t-12}$
	PM1	One-month return reversal
	*PM*71	$Price_{t-1}/Price_{t-7}$
	STDEV	Standard Deviation of daily returns
Composite Models	CTEF	Equal weighted FEP1, FEP2, BR1, BR2, RV1 and RV2
	LAR	Least angular regression
	LRR	Latent root regression
	WLAR	Weighted LAR
	WLRR	Weighted latent regression

This chapter will enable the reader to interpret portfolio performance measures more scientifically. We have tried 36 hypotheses (models) to find out the best model for the test period, see Table 6.1. This process is called data mining. It could be good or bad data mining. And it is not avoidable because no one is going to just try one hypothesis (model) and invest money with it. Through this data mining process, we hope to discover truth, and nothing but the truth. Unfortunately, any statistical discovery comes with skepticism.

Does the truth hold for new data? In our case the question is: will our expectation model make superior money for our clients in the future?

There are two broad categories of testing to add confidence to our judgment of the reported performance. The first one is to adjust each model's performance measurement's statistical significance (p-values) under the context of multiple hypothesis testing. By doing so, we can control the false discovery rate and reduce the chance of making type I errors, i.e., rejecting a true null hypothesis (an investment model has no stock selection ability). This adjustment of the p-value has been applied routinely in medication discovery research. We will detail the adjustment procedure in the next section. The second approach is the empirical Bayes method developed by Markowitz-Xu (1994). This approach is built on the theoretical results of the Bayesian estimation method with the prior assumption that all these methods are equally good. Empirical Bayes estimated the unknown parameters used in the Bayesian method, see Efron and Morris (1973). We use the analysis of variance (ANOVA) method to test the null hypothesis that all these investment models are equally good.

6.2 Single Performance Measurement and Testing

Performance measurements like the Treynor Index, Jensen's alpha, and Sharpe ratio were introduced in Chapter 5. These measurements are point estimates. We will study the estimation error associated with these measurements to get a range of estimates, which enables us to test against the null hypothesis. The statistic of the Sharpe Ratio is from Lo (2002). The statistic of the Treynor index is new material and is derived with the generalized methods of moments (GMM) estimation technique of Hansen (1982).

Assume that we have found an attractive investment model generating portfolio P. The portfolio's geometric mean can be approximated in terms of mean and variance (equation (7), Chapter 5):

$$gm \approx \mu - 0.5 \bullet \sigma^2 \qquad (1)$$

where (μ, σ) are the portfolio's expected return and standard deviation. Let $R_{p,t}$ be the portfolio's return at period t less or equal to total period T. The estimation of (μ, σ) from sample returns is

$$\hat{\mu} = \frac{1}{T}\sum_{t=1}^{T} R_{p,t}$$

$$\hat{\sigma}^2 = \frac{1}{T-1}\sum_{t=1}^{T} (R_{p,t} - \hat{\mu})^2$$

The estimated geometric mean is

$$\widehat{gm} = \hat{\mu} - 0.5 \bullet \hat{\sigma}^2$$

It is easy to check that $\hat{\mu}$ and $\hat{\sigma}^2$ are not correlated. By the property of chi-squared distribution (see section 6.3), the variance of $\hat{\mu}$ and $\hat{\sigma}^2$ are:

$$\text{var}(\hat{\mu}) = \sigma^2 / T$$

and

$$\text{var}(\hat{\sigma}^2) = 2\sigma^4 / (T-1)$$

These equations imply that

$$\sigma_{\widehat{gm}} \sim \sqrt{\text{var}(\hat{\mu}) + 0.5 \cdot 0.5\,\text{var}(\hat{\sigma}^2)} \approx \frac{\sigma\sqrt{1+0.5\sigma^2}}{\sqrt{T}} \approx \frac{\hat{\sigma}\sqrt{1+0.5\hat{\sigma}^2}}{\sqrt{T}} \qquad (2)$$

Geometric mean could be considered as a risk-adjusted measure of performance. The Sharpe ratio is the most popular risk-adjusted performance measure defined as (equation (46), Chapter 5)

$$SR = \frac{\mu - R_f}{\sigma} \tag{3}$$

where R_f is the risk-free rate. Most portfolios' annual Sharpe ratios range from 0 to 1. Having a ratio of one is very good. The estimated Sharpe Ratio is

$$\widehat{SR} = \frac{\widehat{\mu} - R_f}{\widehat{\sigma}}$$

Lo (2002) use GMM technique to derive the estimation error

$$\sqrt{T}(\widehat{SR} - SR) \approx N(0, V_{IID}),$$

where

$$V_{IID} = 1 + \frac{(\mu - R_f)^2}{2\sigma^2} = 1 + 0.5 * SR^2 \tag{4}$$

Lo also derives the formula for the Sharpe ratio and its estimation error in the presence of serial correlations in the returns.

Other performance measurements are measured against benchmarks. It is natural to compare the portfolio's return versus benchmark's return. Let $R_{b,t}$ be the benchmark return at period t less than total period T. Assume that the return difference $R_{p,t} - R_{b,t}$ is i.i.d with normal distribution $N(\mu, \sigma^2)$. The sample mean and standard deviation of return differences are

$$\hat{\mu} = \frac{1}{T}\sum_{t=1}^{T}(R_{p,t} - R_{b,t})$$

$$\hat{\sigma} = \sqrt{\frac{1}{T-1}\sum_{t=1}^{T}(R_{p,t} - R_{b,t} - \hat{\mu})^2}$$

The standard t-statistic is

$$t = \frac{\sqrt{T} * \hat{\mu}}{\hat{\sigma}} \tag{5}$$

which can be used to test the null hypothesis $\mu = 0$.

We report the t-statistics and p-values of excess return for some investment models in Table 6.2.

Table 6.2: Monthly Excess Returns for March 2002 to May 2015

	Average	Standard Dev	t-value	p-value
RLAR	0.79	2.50	3.97	0.00011
RWLAR	0.80	2.55	3.96	0.00011
WLRR	0.75	2.60	3.65	0.00035
CTEF	0.65	2.41	3.41	0.00084
RLASSO	0.67	2.58	3.29	0.00125
SP	0.69	3.07	2.85	0.00495
EP	0.54	2.96	2.29	0.02327
CTEFROIC	0.36	2.66	1.70	0.09038

	Average	Standard Dev	t-value	p-value
BR1	0.12	0.93	1.65	0.10113
ROE1YR	0.18	1.57	1.46	0.14724
FEP1	0.41	3.64	1.41	0.16047
FEP2	0.39	3.66	1.34	0.18241
ROIC	0.17	1.82	1.19	0.23677
PMTREND	0.21	2.47	1.06	0.29017
ALPHA	0.28	3.39	1.06	0.29066
CP	0.26	3.14	1.05	0.29724
PM71	0.25	3.06	1.04	0.30172
BR2	0.08	0.95	1.03	0.30474
CSR	0.14	1.96	0.92	0.35650
ES	0.17	2.34	0.91	0.36490
NCSR	0.12	1.88	0.84	0.40417
ROA1YR	0.10	1.82	0.73	0.46907
NDR	0.12	2.40	0.66	0.51122
ROE3YR	0.07	1.65	0.51	0.60786

Table 6.2 show that the excess return generated by WLRR is highly statistically significant at the 0.035 percent level.

Jensen's alpha is the regression intercept of portfolio excess returns against market excess returns (equation (47), Chapter 5).

$$R_{p,t} - r_{f,t} = \alpha_p + \beta_p (R_{m,t} - r_{f,t}) + \varepsilon_t$$

The t-statistic from the standard regression analysis is

$$t = \frac{\tilde{\alpha}}{std.err(\tilde{\alpha})} \tag{6}$$

which can be used to test the null hypothesis

$$\alpha_p = 0$$

The earliest market risk adjusted performance measurement is the Treynor index (equation (45), Chapter 5)

$$TI = \frac{(\mu - r_f)}{\beta} = \frac{(\mu - r_f) * v_m}{c_{pm}}$$

where v_m is the variance of the market return and c_{pm} is the covariance of the portfolio return with the market return. Denote $\theta \equiv (\mu - r_f, \mu_m - r_f, v_p, v_m, c_{pm})'$ the vector of parameters to be estimated, where the additional parameter μ_m is the market's expected returns and v_p is the portfolio variance σ^2. The moments used in GMM are the means and variances and covariance for the portfolio and market returns. They are defined as

$$\varphi_1(x_t, \theta) = R_{p,t} - \mu$$

$$\varphi_2(x_t, \theta) = R_{m,t} - \mu_m$$

$$\varphi_3(x_t, \theta) = (R_{p,t} - \mu)^2 - v_p$$

$$\varphi_4(x_t,\theta) = (R_{m,t} - \mu_m)^2 - v_m$$

$$\varphi_5(x_t,\theta) = (R_{p,t} - \mu)(R_{m,t} - \mu_m) - c_{pm}$$

Denote moments function vector $\varphi = (\varphi_1,\dots,\varphi_5)'$ and return vector $x_t = (R_{p,t}, R_{m,t})'$. The GMM estimator of θ, denoted by $\hat{\theta}$, is given by the solution to moments conditions

$$\frac{1}{T}\sum_{t=1}^{T}\varphi(x_t,\hat{\theta}) = 0 \tag{7}$$

Hansen (1982) shows that

$$\sqrt{T}(\hat{\theta}-\theta) \sim N(0,V_\theta), V_\theta \equiv H^{-1}\Sigma H^{-1} \tag{8a}$$

where

$$H \equiv \lim_{T\to\infty} E[\frac{1}{T}\sum_{t=1}^{T}\varphi_\theta(X_t,\theta)] \tag{8b}$$

and

$$\Sigma \equiv \lim_{T\to\infty} E[\frac{1}{T}\sum_{t=1}^{T}\sum_{s=1}^{T}\varphi(X_t,\theta)\varphi'(X_s,\theta)] \tag{8c}$$

It turns out that H is a very simple matrix and Σ is easy to calculate. By the definition of φ,

$$\varphi_\theta(X_t,\theta) = \begin{pmatrix} -1 & 0 & 0 & 0 & 0 \\ 0 & -1 & 0 & 0 & 0 \\ -2(R_{p,t}-\mu) & 0 & -1 & 0 & 0 \\ 0 & -2(R_{m,t}-\mu_m) & 0 & -1 & 0 \\ R_{m,t}-\mu_m & R_{p,t}-\mu & 0 & 0 & -1 \end{pmatrix}$$

This implies that

$$H = -I \tag{9}$$

The reader can check that

$$\Sigma = \begin{pmatrix} v_p & c_{pm} & & 0 & \\ c_{pm} & v_m & & & \\ & & 2v_p^2 & 2c_{pm}^2 & 2v_p c_{pm} \\ 0 & & 2c_{pm}^2 & 2v_m^2 & 2v_m c_{pm} \\ & & 2v_p c_{pm} & 2v_m c_{pm} & v_p v_m + c_{pm}^2 \end{pmatrix} \tag{10}$$

Any function of parameter vector θ, $g(\theta)$, can be estimated by the delta method:

$$\sqrt{T}(g(\hat{\theta})-g(\theta)) \sim N(0,V_g), V_g \equiv \frac{\delta g}{\delta\theta}\Sigma\frac{\delta g}{\delta\theta'} \tag{11}$$

The H matrix is eliminated in equation (11) because of equation (9). By applying the delta method to the Treynor index, we have

$$\frac{\delta g}{\delta\theta} = (\frac{v_m}{c_{pm}},\ 0,\ 0,\ \frac{\mu_p - r_f}{c_{pm}},\ -\frac{(\mu_p - r_f)v_m}{c_{pm}^2})$$

A tedious calculation shows

$$V_g = \frac{v_p v_m^2}{c_{pm}^2} + \frac{(\mu_p - r_f)^2 v_m^2}{c_{pm}^2}\left(\frac{v_p v_m}{c_{pm}^2} - 1\right) = \frac{v_m}{\rho^2} + (TI)^2\left(\frac{1}{\rho^2} - 1\right) \tag{12}$$

where $\rho = \frac{c_{pm}}{\sqrt{v_p v_m}}$ is the correlation coefficient between the portfolio and market returns.

6.3 Multiple Hypothesis Testing and False Discovery Rate

In the previous section, we described the most used performance measurements and their error estimations in portfolio management. The error estimation leads us to calculate t-statistics, infer the p-value, which gives us the statistical confidence of assessing the performance of each model. We listed the statistic t_i in Table 6.1 and the corresponding p-value p_i for each investment model using excess return the as performance measurement. The robust least angular regression (RLAR) has the highest t-statistic among the models listed with $t_{\max} = \max\{t_1, t_2, ..., t_n\} = 3.97$ and with p-value 0.00011. We can reject the null hypothesis that RLAR has no excess return. The error of rejecting this hypothesis is less than 0.01 percent. The maximum t-statistic t_{RLAR} is inflated by chance, obviously. If t_i are i.i.d., then

$$P(t_{\max} > t) = 1 - P(t_{\max} \le t) = 1 - \prod_{i=1}^{n} P(t_i \le t) = 1 - \prod_{i=1}^{n} [1 - P(t_i > t)]$$

If we choose t such that $P(t_i > t) = \alpha$,i.e., we control the type I error to be less than or equal to α for each model, the type I error for the luckiest model is

$$P(t_{\max} > t) = 1 - \prod_{i=1}^{n} [1 - P(t_i > t)] = 1 - (1 - \alpha)^n = \overline{\alpha}$$

In case α = 0.05 with the corresponding t = 2 and n = 36, $P(t_{\max} > t) = 0.84$

i.e., the probability of a type I error is almost 85 percent. To have $\overline{\alpha}$ less than 5 percent, we need to choose α to be 0.000145 and the corresponding t-statistic must be at least 3.19.

The adjustment of the t-statistic and α-value arise naturally from multiple hypothesis testing and it is necessary. Bonferroni's (1936) adjustment (equation (17)) is conservative and does not depend on independency assumptions. Under this adjustment the familywise error rate (FWER), which is the probability of making one erroneous rejection of the true null hypothesis, can be controlled at the desired level. The familywise error rate is considered too conservative when the total test methods n is large. The alternative is to control the false discovery rate defined as below.

The general setting for multiple hypothesis is shown in the following grid.

	Called Not Significant	Called Significant	Total
H_0 True	U	V	M_0
H_0 False	T	S	M_1
Total	M-R	R	M

M is the total number of hypotheses tested

M_0 is the number of true null hypotheses, an unknown parameter

M_1 is the number of true alternative hypotheses, an unknown parameter

V is the number of false positives (Type I error) (also called "false discoveries")

U is the number of true positives (also called "true discoveries")

T is the number of false negatives (Type II error)

S is the number of true negatives

R is the number of rejected null hypotheses (also called "discoveries", either true or false)

In M hypothesis tests of which M_0 are true null hypotheses, R is an observable random variable, and V, U, S, and T are unobservable random variables.

The false discovery rate is defined as

$$FDR = E(\frac{V}{R} \mid R > 0) \tag{13}$$

which is the expected portion of the incorrectly rejected null hypotheses among all the rejected null hypotheses. The familywise error rate is

$$FWER = \Pr ob(V > 0) \tag{14}$$

Benjamini and Hochberg (1995) developed the Benjamini, Hochberg, and Yekutieli (BHY) procedure to adjust each individual p-value upward to make it harder to reject the null hypothesis, and the overall false discovery rate can be controlled to a prescribed level when the tests are done independently. Benjamini and Yekutieli (2001) prove that the BHY procedure works the same way when the tests are correlated. Let us denote the collection of hypotheses as $\{H_1, H_2, ..., H_n\}$, the p-values (same as α value) of testing statistics as $\{p_1, p_2, ..., p_n\}$, and ordered p-values as $p_{(1)} \le p_{(2)} \le ... \le p_{(n)}$, the BHY procedure adjusts the p-values sequentially:

$$p_{(i)}^{BHY} = \begin{cases} p_{(n)} & if \quad i = n \\ \min[p_{(i+1)}^{BHY}, \frac{n \cdot C(n)}{i} p_{(i)}] & if \quad i <= n-1 \end{cases} \tag{15}$$

where $C(n) = \sum_{i=1}^{n} \frac{1}{i}$ if the hypotheses are negatively correlated and $C(n) = 1$ if the hypotheses are independent or positively correlated. We will denote the adjusted-p-value as $p_{(i)}^{BHY}$. The BHY adjustment method is very closely related to Holm's method. The difference is the starting point. Holm's method starts adjusting the lowest p-value first.

$$p_{(i)}^{Holm} = \min[\max_{j<=i}\{(n-j+1) \cdot p_{(j)}\}, 1], i = 1, ..., n \tag{16}$$

There are other popular methods to inflate p-values. The Bonferroni method is

$$p_{(i)}^{Bonferroni} = \min[n \cdot p_{(i)}, 1], i = 1, ..., n \tag{17}$$

It is easy to prove

$$p_{(i)}^{Bonferroni} \ge p_{(i)}^{Holm} \tag{18}$$

which implies Bonferroni's adjustment is more conservative than Holm's.

Table 6.3 is the BHY adjusted p-values under positive correlation and negative correlation assumptions of raw p-values applied to the excess returns of investment models of Table 6.1.

Table 6.3: BHY Adjusted p-values

Investment Model	p-value	p-BHY-Positive Corr	p-BHY-Negative-Corr
RLAR	0.0001	0.0009	0.0036
RWLAR	0.0001	0.0009	0.0036
WLRR	0.0002	0.0019	0.0076
CTEF	0.0004	0.0033	0.0136
RLASSO	0.0006	0.0040	0.0162
SP	0.0025	0.0132	0.0536
EP	0.0116	0.0532	0.2158
CTEFROIC	0.0452	0.1798	0.7296
BR1	0.0506	0.1798	0.7296
ROE1YR	0.0736	0.2334	0.8250
FEP1	0.0802	0.2334	0.8250
FEP2	0.0912	0.2432	0.8250
ROIC	0.1184	0.2709	0.8250
PMTREND	0.1451	0.2709	0.8250
ALPHA	0.1453	0.2709	0.8250
CP	0.1486	0.2709	0.8250
PM71	0.1509	0.2709	0.8250
BR2	0.1524	0.2709	0.8250
CSR	0.1782	0.2919	0.8250
ES	0.1825	0.2919	0.8250
NCSR	0.2021	0.3079	0.8250
ROA1YR	0.2345	0.3411	0.8250
NDR	0.2556	0.3556	0.8250
ROE3YR	0.3039	0.4052	0.8250
FY2RV3	0.5037	0.6333	0.8250
DI	0.5146	0.6333	0.8250
STDEV	0.6468	0.7315	0.8250
ROA3YR	0.6595	0.7315	0.8250
ROE5YR	0.6629	0.7315	0.8250
FY1RV3	0.7207	0.7687	0.8250
PM1	0.7742	0.7991	0.8250
ROA5YR	0.8250	0.8250	0.8250

Table 6.3 shows that the excess returns of investment models RLAR, RWLAR, WLRR, CTEF, RLASSO, and SP are still highly statistically positive even under the more conservative BHY procedure adjustment with negative correlation assumptions.

We can adjust the underlying performance measurement to match the new adjusted p-value. If the performance measurement μ has zero as a natural benchmark, for example, excess return, Jensen's alpha, and specific asset selection, the p-value is calculated as

$$P(|t| > \frac{|\hat{\mu}|}{std(\hat{\mu})}) = p$$

If we assume that the estimation error is unchanged such that $std(\hat{\mu}) = std(\mu)$ and p^{adj} are upward adjusted p-values, we define the new performance measurement μ^{adj} as the solution to

$$P(|t| > \frac{|\mu^{adj}|}{std(\hat{\mu})}) = p^{adj}$$

We call μ^{adj} the data-mining adjusted performance measurement, which satisfies $|\mu^{adj}| \le |\hat{\mu}|$ because of $p^{adj} > p$

Table 6.4: BHY Adjusted Average Excess Returns

	Average	BHY-Positive Corr	Adj. Ratio	BHY-Negative Corr	Adj. Ratio
RLAR	0.785	0.629	0.801	0.539	0.686
RWLAR	0.800	0.641	0.802	0.549	0.687
WLRR	0.751	0.605	0.805	0.504	0.672
CTEF	0.649	0.524	0.807	0.425	0.655
RLASSO	0.670	0.548	0.817	0.440	0.657
SP	0.691	0.544	0.786	0.393	0.568
EP	0.537	0.380	0.709	0.185	0.344
CTEFROIC	0.359	0.193	0.539		
BR1	0.121	0.067	0.557		
ROE1YR	0.181	0.091	0.501		
FEP1	0.405	0.210	0.517		
FEP2	0.388	0.202	0.521		
ROIC	0.171	0.088	0.515		
PMTREND	0.208	0.120	0.576		
ALPHA	0.285	0.164	0.577		
CP	0.260	0.152	0.585		
PM71	0.250	0.148	0.590		
BR2	0.077	0.046	0.594		
CSR	0.143	0.085	0.594		
ES	0.168	0.102	0.604		
NCSR	0.124	0.075	0.601		
ROA1YR	0.104	0.059	0.565		
NDR	0.125	0.070	0.563		
ROE3YR	0.067	0.031	0.467		
FY2RV3	-0.002				
DI	-0.013				
STDEV	-0.063				
ROA3YR	-0.058				
ROE5YR	-0.055				
FY1RV3	-0.104				

	Average	BHY-Positive Corr	Adj. Ratio	BHY-Negative Corr	Adj. Ratio
PM1	-0.159				
ROA5YR	-0.133				

Table 6.4 shows that the excess return should be reduced 20 percent to 60 percent. The reduction of excess returns is smaller for the best models.

Harvey and Liu (2014) calculated BHY adjusted p-values using the returns of one hundred fifty simulated long-short investment strategies. Guerard and Xu (2017) calculated the adjusted p-values of asset selection for thirty-six investment models. The SAS routine to perform the BHY procedure, as well as many other multiple-comparison corrections, is PROC MULTTEST.

Program 6.1: Multiple-Comparison Corrections

```
DATA return_data;
   INPUT investments $ Raw_P;
  cards;
    RLAR            0.0001
    RWLAR           0.0001
    WLRR            0.0002
    CTEF            0.0004
    RLASSO          0.0006
    .               .

    .               .
    .               .
    STDEV           0.6468
    ROA3YR          0.6595
    ROE5YR          0.6629
    FY1RV3          0.7207
    PM1             0.7742
    ROA5YR          0.8250

;

PROC SORT DATA=return_data OUT=sorted_p;
   BY Raw_P;

PROC MULTTEST INPVALUES=sorted_p FDR;
Run;
```

The SAS output is shown in Output 6.1.

Output 6.1: Results from Program 6.1

The SAS System

The Multtest Procedure

P-Value Adjustment Information	
P-Value Adjustment	False Discovery Rate

p-Values		
Test	**Raw**	**False Discovery Rate**
1	0.0001	0.0016
2	0.0001	0.0016
3	0.0002	0.0021
4	0.0004	0.0032
5	0.0006	0.0038
6	0.0025	0.0133
7	0.0116	0.0530
8	0.0452	0.1799
9	0.0506	0.1799
10	0.0736	0.2333
11	0.0802	0.2333
12	0.0912	0.2432
13	0.1184	0.2709
14	0.1451	0.2709
15	0.1453	0.2709
16	0.1486	0.2709
17	0.1509	0.2709
18	0.1524	0.2709
19	0.1782	0.2920
20	0.1825	0.2920
21	0.2021	0.3080
22	0.2345	0.3411
23	0.2556	0.3556
24	0.3039	0.4052
25	0.5037	0.6334
26	0.5146	0.6334
27	0.6468	0.7315
28	0.6595	0.7315
29	0.6629	0.7315
30	0.7207	0.7687
31	0.7742	0.7992
32	0.8250	0.8250

6.4 Multiple Mean Comparison Test with ANOVA

We have shown how to build expected returns and portfolio management models in Chapters 4 and 5. For each investment model $i, 1 \le i \le n$, we denote R_{it} the observed return at period t$(1 \le t \le T)$. The observed return fluctuates. We can think of each observed return as randomly drawn from its own population, i.e.,

$$R_{it} = \mu_i + \varepsilon_{it}, 1 \le i \le n, 1 \le t \le T$$

where μ_i is the model i's population mean and ε_{it} is the normally distributed noise.

It is the usual practice to use the raw t-statistic to test whether each model has a positive population mean as shown in Table 6.1. We can and should use the BHY procedure to adjust the p-value to control the false discovery rate to take into account the fact that we have built and tested more than one model.

Instead of focusing the performance of each model against benchmarks, we can compare the performance among the models. Analysis of variance (ANOVA) is the hypothesis-testing technique used to test the equality of two or more population means by examining the variance of observed measurements. The assumptions of ANOVA are:

1. All populations involved follow a normal distribution
2. All populations have the same variance
3. The samples are randomly selected and independent of one another

The one-way ANOVA involves a series of calculations. Let us define the grand mean of all methods and all time periods as

$$\hat{\bar{R}} = \frac{1}{nT}\sum_{i=1}^{n}\sum_{t=1}^{T} R_{i,t},$$

and define the method mean as

$$\bar{R}_i = \frac{1}{T}\sum_{t=1}^{T} R_{i,t}.$$

The within-group sum of squares is defined as

$$SS_w = \sum_{i=1}^{n}\sum_{t=1}^{T}(R_{i,t} - \bar{R}_i)^2 \qquad (19)$$

which has $f_w = n(T-1)$ degrees of freedoms. The between-groups sum of squares is defined as

$$SS_B = T\sum_{i=1}^{n}(\bar{R}_i - \hat{\bar{R}})^2 \qquad (20)$$

which has $f_B = n-1$ degrees of freedom. The degrees of freedom adjusted between-groups and within-group sum of squares are

$$MS_w = SS_w / f_w$$

$$MS_B = SS_B / f_B$$

The F statistic used to test the null hypothesis that all methods' means are equal

$$N_0 : \mu_1 = \mu_2 =, \ldots, = \mu_n$$

is

$$F(f_B, f_w) = \frac{MS_B}{MS_w}$$

The critical value is the number that the test statistic must exceed to reject the null hypothesis. It depends on the number of models and number of observations. It turns out proving $F(f_B, f_w)$ is F-distributed with $(n-1, n(T-1))$ degrees of freedom is quite involved even though we have used this F test routinely. It depends on Cochran's (1934) theorem, which studies the distribution of quadratic expressions of random variables such as SS_w and SS_B.

Theorem (Cochran): Let $X_1,\ldots,X_n$ be independent $N(0,\sigma^2)$ distributed random variables and suppose that the total sum of squares of X_i can be expressed through sum of other quadratic forms,

$$\sum_{i=1}^{n} X_i^2 = \sum_{i=1}^{k} Q_i \tag{21}$$

where $Q_1, Q_2, \ldots, Q_K$ are positive semi-definite quadratic forms of the random variables $X_1,\ldots,X_n$, that is, $Q_i = X'\Lambda_i X$, $i = 1, 2, \ldots, k$. Set $r_i = rank(\Lambda_i)$. If $r_1 + r_2 +, \ldots, + r_k = n$, then $Q_1, Q_2, \ldots, Q_K$ are independently distributed with $Q_i \sim \sigma^2 \chi^2(r_i)$.

Corollary 1: Sum of squares $\sum_{i=1}^{n} X_i^2$ is chi-squared distributed with distribution $\sigma^2 \chi^2(n)$

This follows by rewriting the sum of squares of equation (21) in matrix form

$$\sum_{i=1}^{n} X_i^2 = X' I X$$

where I is the identity matrix with rank n.

Corollary 2: Sum of squares $\sum_{i=1}^{n} (X_i - \bar{X})^2$ is chi-squared distributed with $\sigma^2 \chi^2(n-1)$

Corollary 2 follows from the algebra equations

$$\sum_{i=1}^{n} X_i^2 = \sum_{i=1}^{n} (X_i - \bar{X})^2 + n\bar{X}^2 \tag{22}$$

and

$$n\bar{X}^2 = \frac{1}{n} \sum_{i=1}^{n} \sum_{j=1}^{n} X_i X_j = X' \begin{pmatrix} \frac{1}{n} & .. & \frac{1}{n} \\ .. & .. & .. \\ \frac{1}{n} & .. & \frac{1}{n} \end{pmatrix} X$$

Corollary 2 holds for any i.i.d with $X_i \sim N(\mu, \sigma^2)$. This can be seen by applying the equation (22) to the demeaned variable $X_i - \mu$.

$$\sum_{i=1}^{n} (X_i - \mu)^2 = \sum_{i=1}^{n} ((X_i - \mu) - (\bar{X} - \mu))^2 + n(\bar{X} - \mu)^2 = \sum_{i=1}^{n} (X_i - \bar{X})^2 + n(\bar{X} - \mu)^2$$

Therefore, the sum of squares $\sum_{i=1}^{n}(X_i - \bar{X})^2$ is chi-squared distributed with $\sigma^2\chi^2(n-1)$

Now we can proceed with the F-distribution proof. Under the null hypothesis that all means are equal, $\bar{R}_i$ is independently distributed with $N(\mu, \sigma^2 / T)$ for i = 1,…,n. Corollary 2 implies that SS_B is chi-squared distributed with n-1 degrees of freedom, i.e., $SS_B \sim \sigma^2\chi^2(n-1)$. Corollary 2 also implies $\sum_{t=1}^{T}(R_{it} - \bar{R}_i)^2 \sim \sigma^2\chi^2(T-1)$ a chi-squared distributed with (T-1) degrees of freedom for every *i*. By the assumption (3) of ANOVA, $R_{it} - \bar{R}_i$ is independent of $R_{js} - \bar{R}_j$ if $i \neq j$. This implies $\sum_{t=1}^{T}(R_{it} - \bar{R}_i)^2$ is independent of $\sum_{t=1}^{T}(R_{jt} - \bar{R}_j)^2$ if $i \neq$ j. Therefore, by corollary 1, $SS_w \sim \sigma^2\chi^2(n(T-1))$; chi-squared distributed with n(T-1) degrees of freedom. This proves that $F(f_B, f_w)$ is indeed F-distributed with $(n-1, n(T-1))$ degrees of freedom.

We can apply Cochran's theorem directly to prove that SS_w is chi-squared distributed, which involves stacking n $T \times T$ matrices together. The interested reader can write out Λ_1 and Λ_2. Matrix tensor product notation will help a lot. The SAS procedure for ANOVA is PROC ANOVA.

Program 6.2 shows the one-way ANOVA SAS code and output.

Program 6.2: PROC ANOVA

```
data WORK.DATA1;
      %let _EFIERR_ = 0;
infile 'C:\Sas_one_way_anova_returns.csv' delimiter = ',' MISSOVER DSD
lrecl=32767 firstobs=2;
         informat Investment_Model $20.;
         informat M_Return best32.;
         format Investment_Model $20.;
         format M_Return best12.;
      input
                  Investment_Model$
                  M_Return
      ;
      if _ERROR_ then call symputx('_EFIERR_',1);
run;
proc anova data = Data1;
      class investment_model;
      model m_return = investment_model;
run;
```

There are two parts in this code. The first part reads a comma separated file with investment models and excess returns. The second part runs the one-way ANOVA analysis. The SAS output is shown in Output 6.2.

Output 6.2: One-way ANOVA

The SAS System

The ANOVA Procedure

Class Level Information

Class	Levels	Values
Investment_Model	41	BP BR1 BR2 CP CSI CSR CTEF CTEFROIC DI DP DR EP ES FEP1 FEP2 MCMALPHA MQ NCSR NDR OCFROIC PM1 PM71 PMTREND RBP RCP RDP REP ROA_1YR ROA_3YR ROA_5YR ROE_1YR ROE_3YR ROE_5YR ROIC RSP RV1 RV2 SP STD WLRR_10 WLRR_15

Number of Observations Read	6396
Number of Observations Used	6396

The SAS System

The ANOVA Procedure

Dependent Variable: M_Return

Source	DF	Sum of Squares	Mean Square	F Value	Pr > F
Model	40	325.73060	8.14327	1.45	0.0344
Error	6355	35798.96907	5.63320		
Corrected Total	6395	36124.69967			

R-Square	Coeff Var	Root MSE	M_Return Mean
0.009017	641.3526	2.373436	0.370067

Source	DF	Anova SS	Mean Square	F Value	Pr > F
Investment_Model	40	325.7306012	8.1432650	1.45	0.0344

The F-value is 1.45 with p-value 0.038. This implies that some investment models are genuinely better than other investment models.

ANOVA can be viewed as special case of regression analysis with structured independent variables, whose matrix representation X has blocks of matrices filled with zeros and ones. In this view, testing the equality of all means is testing constraints of βs in the linear regression. A regular normal linear regression model having n observations and p independent variables can be written as

$$y = X\beta + \varepsilon \tag{23}$$

Suppose that there are m linear equality restrictions on the $p \times 1$ parameter vector β as

$$R\beta = r$$

where R is an $m \times p$ matrix of rank m and r is an $m \times 1$ vector, both consisting of known non-stochastic numbers. We can treat constraint $R\beta = r$ as the null hypothesis H_0 . The test-statistics can be calculated in terms of the sum of squares of regression residual SSE without restrictions and SSE_R with restrictions,

$$F = \frac{(SSE_R - SSE)/m}{SSE/(n-p-1)} \sim F(m, n-k-1) \tag{24}$$

In the ANOVA case, the number of observations is $n \times T$, the number of independent variables is n, and the number of constraints is n–1. There are many ways to specify the constraints of our null hypothesis.

The ANOVA analysis assumes the noise terms $\varepsilon_{i,t}$ are i.i.ds. In the case $\varepsilon_t = (\varepsilon_{1,t}, \quad \varepsilon_{2,t}, \quad \dots \quad ,\varepsilon_{n,t})'$ having heteroscedasticity variance-covariance matrix $C_{n\times n}$, we can write the regression equation (23) as

$$C^{-1/2} y_t = C^{-1/2}\mu + C^{-1/2}\varepsilon_t = C^{-1/2}\mu + \eta_t, 1 \le t \le T \tag{25}$$

where transformed noise term η_{it} is i.i.d with normal distribution $N(0,1)$, y_t and μ are n dimensional vectors. The location vector estimation of unrestricted regression is the sample means

$$\hat{\mu}_{n\times 1} = \hat{y}_{n\times 1} \tag{26}$$

The SSE of unrestricted transformed regression is

$$SSE = \sum_{t=1}^{T} (y_t - \hat{y})' C^{-1} (y_t - \hat{y}) = \sum_{t=1}^{T} y_t' C^{-1} y_t' - T\hat{y} C^{-1} \hat{y} \tag{27}$$

The location vector of restricted transformed regression is

$$\hat{\mu}_{n\times 1} = \frac{l' C^{-1} \hat{y}}{l' C^{-1} l} \hat{y}_{n\times 1} \tag{28}$$

The SSE of restricted transform regression is

$$SSE_R = \sum_{t=1}^{T} (y_t - \frac{l' C^{-1} \hat{y}}{l' C^{-1} l} \hat{y})' C^{-1} (y_t - \frac{l' C^{-1} \hat{y}}{l' C^{-1} l} \hat{y}) = \sum_{t=1}^{T} y_t' C^{-1} y_t' - T \frac{(l' C^{-1} \hat{y})^2}{l' C^{-1} l} \tag{29}$$

The F Test statistic according to

$$F = \frac{(SSE_R - SSE)/(n-1)}{SSE/(nT-(n-1))} = \frac{T(\hat{y} C^{-1} \hat{y} - \frac{(l' C^{-1} \hat{y})^2}{l' C^{-1} l})}{SSE/(nT-n-1)} \sim F(n-1, nT-(n-1)) \tag{30}$$

Notice that the above F statistic is calculated based on known variance-covariance structure C. In real life the covariance matrix C is estimated with sample returns. Rao (1959) shows that equation (30) can be simplified as

$$F = \frac{T-n+1}{n-1}\frac{T}{T-1}\left(\sum_{i=1}^{n}\sum_{j=1}^{n}\overline{c}^{ij}\overline{r}_i \cdot \overline{r}_j - \frac{[\sum_{i=1}^{n}\sum_{j=1}^{n}\overline{c}^{ij}(\overline{r}_i + \overline{r}_j)]^2}{4\sum_{i=1}^{n}\sum_{j=1}^{n}\overline{c}^{ij}} \right) \quad (31)$$

where $(\overline{c}^{ij})$ is the inverse matrix of the sample dispersion matrix $\overline{C}$. Miller, Xu, and Guerard (2013) applied the Rao test to 17 models and reported F statistics F=1.9 and rejected the null hypothesis that all models have the same geometric mean with 95 percent of confidence.

6.5 Regression to the Mean

Almost all mutual fund sales prospectuses or other investment disclosures contain this disclaimer: Past performance is no guarantee of future results. It seems odd for an asset manager's sponsor to put up such a disclaimer. After all, the better the past performance or simulated performance, the more likely you would invest in the product. The fundamental issue here is that past performance is the combination of luck and skill, and it is very hard to filter out the luck part. By putting up this disclaimer, the sponsor shifts the responsibility of decoupling the luck and skill to the investor.

Mutual fund performance is very similar to a multiple-choice test score, which is a combination of skill and luck. For students who score above average on any given test, there will be a group who were skilled and did not have any unlucky mishaps, while another group would be unskilled, but extremely lucky. If these same two groups were retested on the same material, the skilled students might have an unlucky mishap, while the unskilled are unlikely to be as lucky as the first time. Hence, those students are unlikely to do quite as well in the second test and some other students who made poor scores on the first test will perform better on the second test. This phenomenon is called "regression to the mean." In the late 19th century, it was observed and popularized in a study of human height by Sir Francis Galton with the publication of *Regression Towards Mediocrity in Hereditary Stature.*

Regression to the mean implies that the best estimate of sample means should shrink to a grand mean. This shrinkage property can be derived as a natural consequence of the Bayesian estimation method, which assumes the sample means themselves has a prior distribution with a common mean. Stein (1956) showed that shrinking the sample means to the grand mean is a "better" method in the context of estimating the multiple means. Stein (1961) formalizes the definition of "better" as a quadratic loss function over the unknown multiple mean space and shows that the James-Stein's shrinkage to the mean is universally better than sample means if the number of estimations is greater than 2.

The practical mean-variance portfolio optimization (see the section titled "A General Form of Portfolio Optimization" in Chapter 5) is a field to test shrinkage estimation method since all the means of security returns and variance-covariance matrix is estimated. Chopra and Ziemba (1993) show that estimation error, in particular the estimation error of the mean, tends to have a big impact on the portfolio combination. On the expected return estimation, Jorion (1991) and Grauer and Hakansson (1995) show the James-Stein shrinkage estimator of means of security returns produces more efficient out-of-sample asset class portfolios than historical sample means. Black and Litterman (1991, 1992) used the Bayesian approach to combine analysts' view of expected return with market-expected return implied by CAPM. On the variance-covariance estimation side, Vasicek (1973) shrank the sample historical betas to get a better estimation of the Capital Asset Pricing Model (CAPM)'s beta as defined by equation (17) in Chapter 5. See also the section titled "Historical Beta" in Chapter 5. Ledoit and Wolf (2003) show the shrinkage estimated covariance matrix as an input to Markowitz's mean-variance optimization produces a more efficient portfolio than the historical sample covariance matrix estimated by using 60 monthly historical returns.

In the first chapter of this book, we argued that the geometric mean is the single best number to characterize an investment method's performance. It is natural to take the logarithm of one plus return since the geometric means are not additive in the sense that two periods' geometric mean is not the sum of one period geometric means. Let R_{it} denote the observed return of method i at period t and define $g_{it} = \ln(1+R_{it})$. We assume that each observed g_{it} consists of two parts

$$g_{it} = \mu_i + \varepsilon_{it} \tag{32}$$

The first term μ_i is investment model i's measurement of skills. It is latent, unobservable, and composed of random variables. The second term ε_{it} is purely random noise. Assume investment model's skills μ_i are randomly drawn from the same population Θ with mean μ_g and standard deviation σ_μ, i.e. $\mu_i, i = 1,\ldots,n$ are identically distributed with mean μ_g and standard deviation σ_μ, where μ_g and σ_μ are scalar variables. Assume that ε_{it} is time serial independent and distributed with mean zeros and variance-covariance matrix Σ. The vector notations for the investment method's geometric mean are $\mu' = (\mu_1,\ldots,\mu_n)$, with observations $g_t' = (g_{1,t},\ldots,g_{n,t})$, and error terms $\varepsilon_t' = (\varepsilon_{1,t},\ldots,\varepsilon_{n,t})$. The assumptions on the underlying investment model's skill are

$$E(\mu) = \mu_g e \tag{33a}$$

$$E[(\mu - \mu_g e)(\mu - \mu_g e)'] = \sigma_\mu^2 I_{nn} \tag{33b}$$

where $e' = (1,\ldots,1)$ is the n dimensional vector of ones. The assumptions for the error terms are

$$E(\varepsilon_t) = 0_n \tag{34a}$$

$$E(\varepsilon_s \varepsilon_t') = \begin{cases} \Sigma, t = s \\ 0_{nn}, s \neq t \end{cases} \tag{34b}$$

$$E(\mu \varepsilon_t') = 0_{nn}, t = 1,..,T \tag{34c}$$

where 0_n is the n-dimensional vector of zeros, and 0_{nn} is a $n \times n$ dimensional matrix of zeros.

The assumption that μ_i be i.i.d with $N(\mu_g, \sigma_\mu^2)$ could be treated, as we will show, as a prior distribution of μ_i in the Bayesian approach. Is the assumption that all these investments models are no different from each other a bad assumption? We don't think so. The stock returns selected by monkeys throwing darts versus human experts, as reported by the Wall Street Journal, are not statistically different. The contest results are measured every six months. So far there is not enough data to shift our prior beliefs.

The random effect model is necessary. Otherwise, sample means are best linear unbiased estimate (BLUE) and maximum likelihood estimate (MLE), and we know that the sample means are not the best estimator in the multiple experiments context. We will show that the BLUE and MLE are the same for the random effect model and display the James-Stein type shrinkage property. A heuristic argument for estimation in the random-effect model is to think of μ_i as a dependent variable and $g_{i.}$ as independent variables. The linear least square estimate of μ_i given $g_{i.}$ is by regression

$$\mu_i = E(\mu_i) + \beta \bullet [g_{i.} - E(g_{i.})] \tag{35a}$$

where β is the regression coefficient of μ_i against $g_{i.}$ which is

$$\beta = \text{cov}(\mu_i, g_{i.}) / \text{var}(g_{i.}) = \frac{\sigma_\mu^2}{\sigma_\mu^2 + \sigma_{\varsigma_{i.}}^2} \qquad (35b)$$

The heuristic estimation, equations (35a) and (35b), clearly shows the shrinkage characteristic. The rest of this section will formalize the heuristic estimate of equations (35a) and (35b).

First, we will proceed to find the BLUE of μ_i for each investment model i. Let us define the unknown weight as $x_t^T = (x_{1,t}, \dots, x_{n,t}), 1 \le t \le T$ plus constant adjustment x_0, a linear estimate is of form

$$\hat{\mu}_i = x_0 + \sum_{t=1}^{T} x_{i,t} g_{i,t} = x_0 + \sum_{t=1}^{T} x_t^{'} g_t \qquad (36)$$

And the loss function is

$$L(x_0, x_1, \dots, x_T) = E(\mu_i - x_0 - \sum_{t=1}^{T} x_t^{'} g_t)^2 \qquad (37)$$

There are n*T + 1 parameters for each investment model. Since the loss function is a quadratic function of unknowns, the minimum is achieved by variables satisfying the first order conditions, which are

$$\frac{\delta L}{\delta x_0} = E\{\mu_i - x_0 - \sum_{s=1}^{T} x_s^{'} g_s\} = 0 \qquad (38a)$$

and

$$\frac{\delta L}{\delta x_t} = E\{g_t * (\mu_i - x_0 - \sum_{s=1}^{T} x_s^{'} g_s)\} = 0_n \qquad (38b)$$

By definition (32) we have

$$E(g_t) = E(\mu + \varepsilon_t) = E(\mu) = \mu_g e$$

Therefore, from equation (38a) we can solve the adjustment constant x_0,

$$x_0 = \mu_g - \mu_g * (\sum_{t=1}^{T} x_t^{'}) e \qquad (39)$$

Substituting x_0 into equation (38b), then equation (38b) becomes

$$E\{g_t * (\mu_i - \mu_g - \sum_{s=1}^{T} x_s^{'} (g_s - \mu_g e)\} = 0_n$$

By definition (32) and the skill and noise independence assumption (34c),

$$E[g_t * (\mu_i - \mu_g)] = E[(\mu + \varepsilon_t)(\mu_i - \mu_g)] = E\mu(\mu_i - \mu_g) = E(\mu(\mu - \mu_g e)' e_i) = \sigma_\mu^2 e_i$$

where $e_i^t = (0, \dots 1, 0, ..0)$ is the vector with the ith component being one and others being zero,

$$E[g_t x_s^{'} (g_s - \mu_g e)] = E[g_t (g_s - \mu_g e)' x_s^{'}] = E[(\mu + \varepsilon_t)(\mu - \mu_g e + \varepsilon_s)' x_s] = [\sigma_\mu^2 I_{nn} + E(\varepsilon_t \varepsilon_s^{'})] x_s$$

Together with serially independence of noise assumption (34b), the first order condition (38b) becomes

$$\sigma_\mu^2 * e_i - (\sum_{s=1}^{T} x_s)\sigma_\mu^2 I - \Sigma * x_t = 0 \tag{40}$$

Summarize equation (40) over t from 1 to T to get

$$T\sigma_\mu^2 * e_i - T(\sum_{s=1}^{T} x_s)\sigma_\mu^2 I - \Sigma * \sum_{s=1}^{T} x_s = 0 \tag{41}$$

Equation (41) gives the solution

$$\sum_{s=1}^{T} x_s = (T\sigma_\mu^2 I + \Sigma)^{-1} T\sigma_\mu^2 * e_i = (\sigma_\mu^2 I + \Sigma / T)^{-1} \sigma_\mu^2 * e_i \tag{42}$$

Equation (40) implies that weight x_t is independent of t. Together with equation (41),

$$x_t = \frac{1}{T}\sum_{s=1}^{T} x_s = \frac{1}{T}(\sigma_\mu^2 I + \Sigma / T)^{-1} \sigma_\mu^2 * e_i \tag{43}$$

Equations (39) and (42) solve the adjustment factor x_0 as

$$x_0 = \mu_g - \mu_g * (\sum_{t=1}^{T} x_t^{'})e = \mu_g - \mu_g \sigma_\mu^2 e_i^{'}(\sigma_\mu^2 I + \Sigma / T)^{-1} e \tag{44}$$

Substituting (44) and (43) into equation (36), we get the best linear estimate

$$\hat{\mu}_i = \mu_g + \sigma_\mu^2 e_i^T (\sigma_\mu^2 I + \frac{\Sigma}{T})^{-1} (\bar{g}_i - \mu_g e) \tag{45}$$

The entire estimate vector is

$$\hat{\mu} = (\hat{\mu}_1, ..., \hat{\mu}_n)' = \mu_g e + \sigma_\mu^2 (\sigma_\mu^2 I + \Sigma / T)^{-1} (\bar{g} - \mu_g e) \tag{46}$$

When the noise terms are i.i.d, the BLUE of equation (46) is the same as the heuristic estimate (35). The best linear estimation of μ given by equation (46) is the same as maximum likelihood estimate under the normality assumption of prior and error terms follows from standard Bayesian analysis, see Zellner (1987). Assume that $N(\mu_g, \sigma_\mu^2 I)$ is the prior distribution of μ , i.e., the density function of μ is

$$\rho(\mu) = \frac{1}{\sqrt{2\pi\sigma_\mu^{2n}}} * \exp\{-0.5 * (\mu - \mu_g * e)^{'} (\sigma_\mu^{\,2} I)^{-1} (\mu - \mu_g * e)\}$$

The conditional density function of g_t given (μ, Σ) is

$$\rho(g_{it}, i = 1, ..., n, t = 1, .., T \mid \mu, \Sigma) = \prod_{t=1}^{T} \frac{1}{\sqrt[T]{2\pi \det(\Sigma)}} * \exp\{-0.5 * (g_t - \mu)^{'} \Sigma^{-1} (g_t - \mu)\}$$

Manipulation of formulas gives us

$$\rho(\mu \mid g_{it}, i = 1, ..., n, t = 1, .., T)$$
$$\propto \rho(\mu) * \rho(g_{it}, i = 1, ..., n, t = 1, .., T \mid \mu, \Sigma)$$

$$= \frac{1}{\sqrt{2\pi\sigma_\mu^{2n}}} * \exp\{-0.5 * (\mu - \mu_g * e)'(\sigma_\mu{}^2 I)^{-1}(\mu - \mu_g * e)\} *$$

$$\prod_{t=1}^{T} \frac{1}{\sqrt[T]{2\pi \det(\Sigma)}} * \exp\{-0.5 * (g_t - \mu)'\Sigma^{-1}(g_t - \mu)\}$$

$$\propto \exp\left\{0.5 \cdot (\mu - \hat{\mu})'(T\Sigma^{-1} + \sigma_\mu^{-2} I)(\mu - \hat{\mu})\right\}$$

Therefore, the maximum likelihood estimates of μ given observation g_{it} is given by equation (46), which is the same as the best linear estimate. Markowitz and Xu (1994) call a performance measurement setup with heteroscedasticity covariance (equations (32), (33), and (34)) data mining correction model III. The shrinkage estimate is given by equation (46) with parameters $(\mu_g, \sigma_\mu^2, \Sigma)$ being replaced by estimated parameters $(\hat{\mu}_g, \hat{\sigma}_\mu^2, \hat{\Sigma})$. The $(i, j)^{th}$ element of covariance matrix $\hat{\Sigma}$ is given

$$\hat{\sigma}_{ij} = \frac{\sum_{t=1}^{T} (g_{it} - g_{i.})(g_{jt} - g_{j.})}{T - 1} \quad (47)$$

Parameters $(\hat{\mu}_g, \hat{\sigma}_\mu^2)$ are estimated as the sample mean and sample variance of $g_{i.}$. We will formalize the estimation in the next section. Table 6.5 reports the sample estimation and Bayesian estimation of 17 investment models from Miller, Xu, and Guerard (2013).

Table 6.5: Data Mining Corrections Tests, Jan 1980–Dec 2009

Portfolio	$\bar{r}_i - \bar{r}$	Bayesian Estimate of $\bar{r}_i - \bar{r}$	Estimate to Actual Ratio
S&P500	-0.09	-0.08	0.96
USER	0.14	0.12	0.86
BR1	0.06	0.05	0.84
BR2	0.03	0.02	0.59
RV1	0.07	0.09	1.18
RV2	-0.10	-0.08	0.82
FEP1	-0.15	-0.09	0.59
FEP2	-0.40	-0.32	0.79
CTEF	0.16	0.16	0.95
EP	-0.05	-0.05	1.05
BP	-0.10	-0.10	1.01
CP	0.02	0.02	0.82
SP	0.18	0.17	0.94
DP	0.09	0.09	0.96
PM71	-0.02	-0.03	1.26
PM71	-0.08	-0.09	1.16
EWC	0.00	-0.01	1.30
MQ	0.24	0.21	0.91

The second column displays each model's performance difference relative to the grand mean. The third column displays the Bayesian's estimated performance difference. The fourth column displays the ratio of the Bayesian's estimate to the model's original performance measurement. In general, the ratios are all close to one except EWC, which is 1.30, and BR2, which is 0.59. The performance reduction of our favorite models USER (WLRR in this book) and CTEF are only 15 percent and 5 percent, which are not a statistically significant.

The shrinkage property of estimation formula (46) is hidden by the matrix expression

$$\sigma_\mu^2(\sigma_\mu^2 I + \Sigma / T)^{-1} \tag{48}$$

We make more assumptions on the covariance structure Σ of the error term ε_{it} based on the fact that the market swing is the dominant force for all investment models. When the market goes up, every model is doing well, and when market goes down every model goes down too. We can model this market effect directly by assuming that

$$g_{it} = \mu_i + z_t + \zeta_{it} \tag{49}$$

where z_t is the market effect and ζ_{it} is i.i.d return noise. Markowitz and Xu call this data mining correction model II. The estimation formula (46) becomes

$$\hat{\mu} = \mu_g e + \frac{\sigma_\mu^2}{\sigma_\mu^2 + \sigma_\varsigma^2 / T}(\bar{g} - \bar{\bar{g}}e) + \frac{\sigma_\mu^2}{\sigma_\mu^2 + \sigma_\varsigma^2 / T} \bullet \frac{\sigma_\mu^2}{\sigma_\mu^2 + \sigma_\varsigma^2 / T + n^* \sigma_z^2 / T}(\bar{\bar{g}} - \mu_g) \tag{50}$$

where $\bar{\bar{g}}$ is the grand mean, i.e., the average of sample means of all models. In practice, we always choose prior mean μ_g of all investment methods as $\bar{\bar{g}}$. The formula (50) is simplified to

$$\hat{\mu} = \bar{\bar{g}}e + \beta(\bar{g} - \bar{\bar{g}}e) \tag{51a}$$

where

$$\beta = \frac{\sigma_\mu^2}{\sigma_\mu^2 + \sigma_\varsigma^2 / T} \tag{51b}$$

The shrinkage property is explicit now because by β is clearly less than one.

We can model the market effect z_t explicitly by defining z_t to be logarithm of one plus market (for example, S&P 500) return. Then model (49) becomes

$$\tilde{g}_{it} = g_{it} - z_t = \mu_i + \zeta_{it} \tag{52}$$

The best of estimator of μ are still given by equations (51). Markowitz-Xu call equation (52) data mining correction model I.

6.6 Empirical Bayes Estimation and Hypothesis Testing

Under the assumption that each investment model's unobservable performance measurement's mean μ_i is a randomly drawn from the same population, the Bayesian's estimate of future mean $\hat{\mu}_i$ shrinks the sample mean $\bar{g}_i$ toward grand sample mean $\bar{\bar{g}}$. The shrinkage factor, β , depends on the variance σ_μ of the population method, is given by equation (51). The empirical Bayes replaces the true parameters $(\sigma_{\mu}^2,\sigma_{\zeta}^2)$ with sample estimated parameters $(\hat{\sigma}_{\mu}^2,\hat{\sigma}_{\zeta}^2)$. The empirical Bayes estimate of β is

$$\hat{\beta}=\frac{\hat{\sigma}_{\mu}^2}{\hat{\sigma}_{\mu}^2+\hat{\sigma}_{\varsigma}^2/T} \tag{51c}$$

The analysis of variance (ANOVA) will provide convenient notations to facilitate the estimation of parameters. Since the term "model" is a reserved key word for SAS and is more commonly used in the statistical world, we switch to use "method" in the place of investment model.

Table 6.6: ANOVA and Sum of Squares

Source	d.f.	SS	MS	F
Method	n-1	$SS_{method}=T\sum_{i=1}^{n}(g_{i.}-g_{..})^2$	$MS_{method}=\frac{SS_{method}}{n-1}$	$\frac{MS_{method}}{MS_E}$
Market	T-1	$SS_{market}=n\sum_{t=1}^{n}(g_{.t}-g_{..})^2$	$MS_{Market}=\frac{SS_{market}}{T-1}$	$\frac{MS_{market}}{MS_E}$
Error	(n-1)(T-1)	$SS_E=\sum_{i,t=1}^{n}(g_{i,t}-g_{i.}-g_{.t}+g_{..})^2$	$MS_E=\frac{SS_E}{(n-1)*(T-1)}$	
Total	nT-1	$SS_{total}=\sum_{i=1}^{n}\sum_{t=1}^{T}(g_{i,t}-g_{..})^2$		

The variables $g_{i.}$, $g_{.t}$, and $g_{..}$ in Table 6.6 are the sample average of given method, sample average of given market period, and sample grand average.

$$g_{i.}=\frac{1}{T}\sum_{t=1}^{T}g_{it} \tag{53a}$$

$$g_{.t}=\frac{1}{n}\sum_{i=1}^{n}g_{it} \tag{53b}$$

$$g_{..}=\frac{1}{nT}\sum_{i=1}^{n}\sum_{t=1}^{T}g_{it}=\frac{1}{n}\sum_{i=1}^{n}g_{i.}=\frac{1}{T}\sum_{t=1}^{T}g_{.t} \tag{53c}$$

In our case, the error term SS_E is the same as $SS_{method,market}$, the interactive term of method and market in the standard two-way analysis. The total sum of squares is the sum of components of the sum of squares

$$SS_{total}=SS_{method}+SS_{market}+SS_E \tag{54}$$

It follows that the observed deviation from the grand mean can be decomposed of deviations with respect to method means and the market mean as

$$g_{i,t}-g_{..}=(g_{i,.}-g_{..})+(g_{.,t}-g_{..})+g_{i,t}-g_{i,.}-g_{.,t}+g_{..} \tag{55}$$

where $g_{i,.} - g_{..}$ is a component of SS_{method}, $g_{.t} - g_{..}$ is a components of SS_{market}, and $g_{i,t} - g_{i,.} - g_{.,t} + g_{..}$ is a component of SS_E as defined in Table 6.5. Taking squares on both sides of equation (55) leads to

$$SS_{total} = SS_{market} + SS_{period} + SS_E$$
$$+2\{\sum_{i=1}^{n}\sum_{t=1}^{T}(g_{i.} - g_{..})(g_{.t} - g_{..}) + \sum_{i=1}^{n}\sum_{t=1}^{T}(g_{i.} - g_{..})(g_{it} - g_{i.} - g_{.t} + g_{..})$$
$$+\sum_{i=1}^{n}\sum_{t=1}^{T}(g_{.t} - g_{..})(g_{it} - g_{i.} - g_{.t} + g_{..})\}$$

The decomposition property follows from the fact that the sums of cross product terms add up to zeros by definition of (53). Now we proceed to link the expected value of sum of squares SS_{method}, SS_{market}, and SS_{total} with the estimated parameters $(\sigma_\mu^2, \sigma_\zeta^2)$. For each component of SS_{method},

$$E(g_{i.} - g_{..})^2 = E(\mu_i - \hat{\mu} + \zeta_{i.} - \zeta_{..})^2 = E(\mu_i - \hat{\mu})^2 + E(\zeta_{i.} - \zeta_{..})^2$$
$$= (1 - \frac{1}{n})(\sigma_\mu^2 + E(\zeta_{i.}^2) = (1 - \frac{1}{n})(\sigma_\mu^2 + \frac{1}{T}\sigma_\zeta^2)$$

which leads to

$$E(SS_{method}) = T\sum_{i=1}^{n} E(g_{i.} - g_{..})^2 = Tn(1 - \frac{1}{n})(\sigma_\mu^2 + \frac{1}{T}\sigma_\zeta^2) = (n-1)T(\sigma_\mu^2 + \frac{1}{T}\sigma_\zeta^2) \quad (56)$$

For each component of SS_{market},

$$E(g_{.t} - g_{..})^2 = E(z_t + \zeta_{.t} - (\hat{z} + \zeta_{..}))^2 = E(z_i - \hat{z})^2 + E(\zeta_{.t} - \zeta_{..})^2$$
$$= (1 - \frac{1}{T})(\sigma_z^2 + E(\zeta_{.t}^2)) = (1 - \frac{1}{T})(\sigma_z^2 + \frac{1}{n}\sigma_\zeta^2)$$

which leads to

$$E(SS_{market}) = n\sum_{t=1}^{T} E(g_{.t} - g_{..})^2 = nT(1 - \frac{1}{T})(\sigma_z^2 + \frac{1}{n}\sigma_\zeta^2) = (T-1)(n\sigma_z^2 + \sigma_\zeta^2) \quad (57)$$

For each component of SS_{total},

$$E(g_{it} - g_{..})^2 = E(\mu_i + z_t + \zeta_{i,t} - (\hat{\mu} + \hat{z} + \zeta_{..}))^2 = E(\mu_i - \hat{\mu})^2 + E(z_t - \hat{z})^2 + E(\zeta_{it} - \zeta_{..})^2$$
$$= (1 - \frac{1}{n})\sigma_\mu^2 + (1 - \frac{1}{T})\sigma_z^2 + (1 - \frac{1}{nT})\sigma_\zeta^2$$

which leads to

$$E(SS_{total}) = T(n-1) * \sigma_\mu^2 + n(T-1) * \sigma_z^2 + (nT-1) * \sigma_\zeta^2 \quad (58)$$

By decomposition property (54) of total sum of squares,

$$E(SS_E) = E(SS_{total}) - E(SS_{market}) - E(SS_{method}) = (T-1)(n-1)\sigma_\zeta^2 \quad (59)$$

The three equations (57), (58), and (59) link the expected values of sum of squares, $E(SS_{method})$, $E(SS_{market})$, and $E(SS_E)$ with three unknown parameters $(\sigma_\mu^2, \sigma_z^2, \sigma_\zeta^2)$. We can invert these three equations to express the parameters in terms of the expectation of sum of squares. From equation (59), we have

$$\sigma_\varsigma^2 = \frac{1}{(T-1)(n-1)} E(SS_E) \tag{60}$$

Equation (57) implies

$$\sigma_\mu^2 + \frac{1}{T}\sigma_\varsigma^2 = \frac{1}{(n-1)T} E(SS_{method}) \tag{61}$$

Together with equation (60),

$$\sigma_\mu^2 = \frac{1}{(n-1)T}[E(SS_{method}) - (n-1)\sigma_\zeta^2) = \frac{1}{(n-1)T}[E(SS_{method}) - \frac{1}{T-1}E(SS_E)] \tag{62}$$

The Bayesian β by equation (51b) in terms of expected value of sum of squares is

$$\begin{aligned} \beta &= \frac{\sigma_\mu^2}{\sigma_\mu^2 + \sigma_\varsigma^2 / T} \\ &= \frac{E(SS_{mtehod}) - E(SS_E)/(T-1))}{E(SS_{method})} = 1 - \frac{E(SS_E)}{(T-1)E(SS_{method})} = 1 - \frac{E(MS_E)}{E(MS_{method})} \end{aligned} \tag{63}$$

The empirical Bayes estimate of shrinkage factor β by equation (51c) is

$$\hat{\beta} = \frac{\hat{\sigma}_\mu^2}{\hat{\sigma}_\mu^2 + \hat{\sigma}_\varsigma^2 / T} = 1 - \frac{MS_E}{MS_{method}} = 1 - \frac{1}{F} \tag{64}$$

It is natural to set $\hat{\beta} = 0$ if $\hat{\beta} < 0$.

With $\hat{\beta} > 0$, it is natural to ask whether it is statistically significant.

Theorem: If σ_μ^2=0, i.e., $\mu_i = \mu_g$ are constant, market effect z and error term ε are normally distributed i.i.d, then

$$F = \frac{MS_{method}}{MS_E} \tag{65}$$

is F(n-1, (n-1)*(T-1)) distributed.

Proof:

We can write

$$g_{i,.} - g_{..} = \mu_i + \zeta_{i,.} - (\hat{\mu} + \zeta_{..})$$

Since $\mu_i + \zeta_{i.}$ are i.i.d with variance $\sigma_\mu^2 + \frac{1}{T}\sigma_\varsigma^2$, Corollary 2 of the Cochran Theorem implies that

$$\sum_{i=1}^{n}(g_{i.} - g_{..})^2 = \sum_{i=1}^{n}(\mu_i + \zeta_{i,.} - (\hat{\mu} + \zeta_{..}))^2$$ is chi-squared distributed with $(\sigma_\mu^2 + \frac{1}{T}\sigma_\varsigma^2)\chi_{(n-1)}^2$.

Therefore,

$$SS_{method} \sim T(\sigma_\mu^2 + \frac{1}{T}\sigma_\zeta^2)\chi_{(n-1)}^2$$

Corollary 2 of Cochran's theorem also implies that

$$SS_E \sim \sigma_\zeta^2 \chi_{(n-1)(T-1)}^2$$

By definition of F-value,

$$F = \frac{MS_{method}}{MS_E} = \frac{SS_{method} / (n-1)}{SS_E / (n-1)(T-1)} \sim \frac{T\sigma_\mu^2 + \sigma_\varsigma^2}{\sigma_\varsigma^2} \frac{\chi_{n-1}^2 / (n-1)}{\chi_{(n-1)(T-1)}^2 / (n-1)(T-1)}$$

If σ_μ equals 0,

$$F = \frac{MS_{method}}{MS_E} \sim \frac{\chi_{n-1}^2 / (n-1)}{\chi_{(n-1)(T-1)}^2 / (n-1)(T-1)}$$

Cochran's theorem also implies the independence MS_{method} of MS_E. By the definition of F-distribution, F-value calculated as (65) is F-distributed with $(n-1, (n-1)(T-1))$ degree of freedoms.

Reader can use the SAS two-way ANOVA procedure to calculate the sum of the squares, F-value of equation (65), and the corresponding p-values as shown in Program 6.3.

Program 6.3: Two-way ANOVA

```
data WORK.DATA2;
       %let _EFIERR_ = 0;
infile 'C:\Sas_two_way_anova_returns.csv' delimiter = ',' MISSOVER DSD
lrecl=32767 firstobs=2;
          informat Investment_Model $20.;
          informat Market best32.;
          informat M_Return best32.;
          format Investment_Model $20.;
          format Market best12.;
          format M_Return best12.;
       input
Investment_Model $
Market
M_Return
       ;
       if _ERROR_ then call symputx('_EFIERR_',1);
run;
proc anova data=Data2;
class Investment_Model Market;
model M_return = Investment_Model Market;
run;
```

The SAS code has two parts: one reads the comma separated file and the other performs the ANOVA analysis. See Output 6.3.

Output 6.3: Results of Program 6.3

The SAS System

The ANOVA Procedure

Class Level Information

Class	Levels	Values
Investment_Model	41	BP BR1 BR2 CP CSI CSR CTEF CTEFROIC DI DP DR EP ES FEP1 FEP2 MCMALPHA MQ NCSR NDR OCFROIC PM1 PM71 PMTREND RBP RCP RDP REP ROA_1YR ROA_3YR ROA_5YR ROE_1YR ROE_3YR ROE_5YR ROIC RSP RV1 RV2 SP STD WLRR_10 WLRR_15
Market	156	200301 200302 200303 200304 200305 200306 200307 200308 200309 200310 200311 200312 200401 200402 200403 200404 200405 200406 200407 200408 200409 200410 200411 200412 200501 200502 200503 200504 200505 200506 200507 200508 200509 200510 200511 200512 200601 200602 200603 200604 200605 200606 200607 200608 200609 200610 200611 200612 200701 200702 200703 200704 200705 200706 200707 200708 200709 200710 200711 200712 200801 200802 200803 200804 200805 200806 200807 200808 200809 200810 200811 200812 200901 200902 200903 200904 200905 200906 200907 200908 200909 200910 200911 200912 201001 201002 201003 201004 201005 201006 201007 201008 201009 201010 201011 201012 201101 201102 201103 201104 201105 201106 201107 201108 201109 201110 201111 201112 201201 201202 201203 201204 201205 201206 201207 201208 201209 201210 201211 201212 201301 201302 201303 201304 201305 201306 201307 201308 201309 201310 201311 201312 201401 201402 201403 201404 201405 201406 201407 201408 201409 201410 201411 201412 201501 201502 201503 201504 201505 201506 201507 201508 201509 201510 201511 201512

Number of Observations Read	6396
Number of Observations Used	6396

The SAS System

The ANOVA Procedure

Dependent Variable: M_Return

Source	DF	Sum of Squares	Mean Square	F Value	Pr > F
Model	195	116800.3767	598.9763	201.10	<.0001

Source	DF	Sum of Squares	Mean Square	F Value	Pr > F
Error	6200	18466.2619	2.9784		
Corrected Total	6395	135266.6386			

R-Square	Coeff Var	Root MSE	M_Return Mean
0.863483	164.4253	1.725813	1.049603

Source	DF	Anova SS	Mean Square	F Value	Pr > F
Investment_Model	40	309.9684	7.7492	2.60	<.0001
Market	155	116490.4083	751.5510	252.33	<.0001

The F-value is 2.6 with p-value less than 0.0001. The two-way ANOVA analysis provides stronger support to rejecting the null hypothesis than the one-way analysis does.

6.7 Summary and Conclusions

We reviewed the most popular performance measures for investment models in the first section of this chapter. Those measures include geometric means, excess returns, Jensen's alpha, the Treynor index, and Sharpe ratio. Since the measures are estimated using sample data, we also presented the estimation error calculation, which enabled us to do a statistical significance test. All the tests were done in the context of a single hypothesis.

We introduced the BHY procedure to control the false discovery rate in the multiple tests. The original purpose of the false discovery rate was to find a few reasonable candidates among thousands and hundreds of thousands of candidates for further investigating. The application of the false discovery rate to financial modeling should be the same in principle, i.e., apply FDR to information coefficients or spread of return of investment models to find a few reasonable models for fully optimized and managed portfolio simulations. The classic ANOVA can be applied to test whether these investment models are truly different. In the last section, we showed that the random-effect model is the unified model to test equality of the models and to correct the sample estimation of each model in the context of data mining.

Chapter 7: Summary and Conclusions

The purpose of this book has been to how readers how use SAS to enhance their wealth.

We disclosed our beliefs at the outset. We believe in Active Quantitative Management using the portfolio selection, construction, and management techniques of Harry Markowitz, William (Bill) Sharpe, Jan Mossin, John Blin, Henry Latane, Martin Gruber, Ned Elton, Barr Rosenberg, Haim Levy, and the investment professionals at Factset, FIS Global, and Axioma. We believe that the empirical evidence of the past 30 years suggests that financial anomalies were identified, have persisted, and most likely will persist into the coming decade. We believe the portfolio benchmarks established by Markowitz, Sharpe, and Blin are still relevant and difficult to beat. We believe that Big Data will enhance returns in the future, but the enhancements will be more in the 15–20 percent range, rather than doubling existing excess returns. We have presented models and updated analyses to show that robust regression, as we estimated the financial models in the early 1990s, still works.

Business people seek to maximize the stock price of their firms. The stock price of a company should be dependent upon its historical earnings and its expectational earnings. An intelligent investor should be prepared to read, understand, and implement an investment strategy that incorporates financial data of firms. An investor buys stock, representing (often fractional) ownership of the firm. An investor can receive dividends paid from the earnings of the firm. An investor pays stock in anticipation that the price of the stock will rise in the future. We show the relevance of financial accounting data to estimate financial ratios. The financial community has long calculated ratios to assess the liquidity, profitability, leverage, and efficiency of firms. Ratio calculations and analysis to summarize financial information can be found on the balance sheet and income statement. Investors can often easily assess the financial health of a firm by calculating the Altman-Z calculation. The Altman Z-statistic is estimated and continues to be relevant. The Altman Z-bankruptcy prediction model and corporate exports are important concepts developed in this text.

Financial services, such as FactSet and Compustat, gather data on corporations for the investment community and produce a similar format for all firms to make it easy to compare companies. Although the balance sheet or position statement is a useful quantitative picture of a firm's financial position, it is not an exact reflection of the firm's economic worth. The balance sheet is constructed on the basis of formal rules and does not necessarily represent the market value of the firm as either a going concern or liquidated (sold off) entirely.

Earnings forecasting work has greatly enhanced portfolio returns in US, and, particularly, non-US markets. Financial models, when properly developed and tested with proper transactions costs work in about 75 percent of the years. The models produce statistically significant excess returns in the years that the models win, but ONLY if the models are used religiously, 100 percent of the time, investors are fully invested, and asset owners fully invest the Mean-Variance weights (or Equal Active Weights, plus or minus the benchmark, at least two percent active weighting). We have shown that robust regression, using the Tukey Optimal Influence Function produces highly statistically significant stock selection models delivering portfolio returns passing the Markowitz- Xu Data Mining Corrections test.

Why do we maximize the Geometric Mean of portfolios? Henry Latane and Harry Markowitz taught us in 1959 that to maximize the Geometric Mean maximizes the utility of final wealth; achieving the greatest level of terminal wealth, in the shortest time possible. Henry Latane remarked that the Efficient Markets hypothesis only said that the average investor only earned an average return, adjusted for risk. We believe that smart people, with good databases, can enhance returns about 1–2 percent, annualized, adjusted for risk and risk premiums accepted (knowing or unknowingly incurred).

The purpose of this book has been to show readers how use SAS to enhance their wealth through choosing a statistically significant stock selection model, creating Mean-Variance efficient portfolios, and be aggressive in investing to maximize the Geometric Mean. The investment professionals at Factset, FIS Global, and Axioma have investment platforms that can implement these techniques. We believe that the empirical evidence of the past 30 years suggests that financial anomalies were identified, have persisted, and most likely will persist into the coming decade.

References

Ackoff, R.L. 1999. Re-Creating the Corporation: a Design of Organizations for the 21st Century, Oxford, UK: Oxford University Press.

Alexander, S.S. 1961. "Price Movements in Speculative Markets: Trends or Random Walks." *Industrial Management Review* 2, 7–26. Reprinted in Cootner (1964).

Altman, E.I. 1968. "Financial Ratios, Discriminant Analysis and the Prediction of Corporate Bankruptcy." *The Journal of Finance* 23, no.4, 589–609.

Andrews, D.F., P.J. Bickel, F.R. Hampel, P.J. Huber, W.H. Rogers, and J.W. Tukey. 1972. *Robust Estimates of Location: Survey and Advances.* Princeton, NJ: Princeton University Press.

Ball, R. and P. Brown. 1968. "An Empirical Evaluation of Accounting Income Numbers." *Journal of Accounting Research* 6, 169–178.

Basu, S. 1977. "Investment Performance of Common Stocks in Relation to their Price-Earnings Ratios: A Test of the Efficient Market Hypothesis." *The Journal of Finance* 32, no.3, 663–682.

Beaton, A.E. and J.W. Tukey. 1974. "The Fitting of Power Series, Meaning Polynomials, Illustrated on Band-Spectroscopic Data." *Technometrics* 16, no.2, 147–185.

Belsley, D.A., E. Kuh, and R.E. Welsch. 1980. *Regression Diagnostics: Identifying Influential Data and Sources of Collinearity.* New York: John Wiley & Sons.

Benjamini, Y. and D. Yekutieli. 2001. "The Control of the False Discovery Rate in Multiple Testing Under Dependency." *The Annals of Statistics* 29, no.4, 1165–1188.

Benjamini, Y., and Y. Hochberg. 1995. "Controlling the false discovery rate: a practical and powerful approach to multiple testing." *Journal of the Royal Statistical Society* B 57: no.1, 289–300

Berk, J.B. and J.H. van Binsbergen. 2016. "Assessing Asset Pricing Models using Revealed Preference." *Journal of Financial Economics* 119, no.1, 1–23.

Berk, J.B. and J.H. van Binsbergen. 2017. "Leveling the Playing Field." In J.B. Guerard, Jr., Ed. *Portfolio Construction, Measurement, and Efficiency: Essays in Honor of Jack Treynor.* New York: Springer.

Bernstein, L.A. 1989. *Financial Statement Analysis: Theory, Application, and Interpretation*, 4th edition, Homewood, IL: Irwin, Chapters 4–11.

Black, F. 1972. "Capital market equilibrium with restricted borrowing." *The Journal of Business* 45, no.3, 444–455.

Black, F., M.C. Jensen, and M. Scholes. 1972. "The Capital Asset Pricing Model: Some Empirical Tests." In *Studies in the Theory of Capital Markets*, edited by M.C. Jensen. New York: Praeger.

Black, Fischer, and R. Litterman. 1991. "Asset Allocation: Combining Investor Views with Market Equilibrium." *The Journal of Fixed Income* 1, no.2, 7–18.

Black, Fischer, and R. Litterman. 1992. "Global Portfolio Optimization." *Financial Analysts Journal* 48(5), 28–43.

Blin, J. S. Bender, and J. B.Guerard, Jr. 1997. "Earnings Forecasts, Revisions, and Momentum in the Estimation of Efficient Market-Neutral Japanese and U.S. Portfolios." In *Research in Finance Volume* 15, edited by A.H. Chen. Bingley, UK: Emerald Group.

Bloch, M., J. Guerard, Jr., H. Markowitz, P. Todd, and G Xu. 1993. "A Comparison of Some Aspects of the U.S. and Japanese Equity Markets." *Japan and the World Economy* 5, no.1, 3–26.

Board of Governors of the Federal Reserve System. 2016. "Financial Accounts of the United States: Flow of Funds, Balance Sheets, and Integrated Macroeconomic Accounts." https://www.federalreserve.gov/releases/z1/current/z1.pdf

Bogle, J.C. 1999. *Common Sense on Mutual Funds: New Imperatives for the Intelligent Investor.* New York: John Wiley & Sons.

Bogle, J.C. 2010. *Common Sense on Mutual Funds.* Fully updated 10th anniversary edition. New York: John Wiley & Sons.

Bonferroni, C. E. 1936. "Teoria statistica delle classi e calcolo delle probabilità." Florence, Italy: Seeber.

Brealey, R.A. and S.C. Myers. 2003. *Principles of Corporate Finance*, 7th edition. New York: McGraw-Hill/Irwin, Chapter 31.

Brealey, R.A., S.C. Myers, and F. Allen. 2006. *Principles of Corporate Finance.* 8th edition. New York: McGraw-Hill/Irwin, Chapter 8.

Brinson, Gary P., B.D. Singer, and G.L. Beebower. 1991. "Determinants of Portfolio Performance II: An Update." *Financial Analysts Journal* 47, no.3, 40–48.

Brown, S. J. 1979 "The Effect of Estimation Risk on Capital Market Equilibrium." Journal of Financial and Quantitative Analysis 14, no. 2, 215–220

Brush, J.S. 2007. "A Flexible Theory of Price Momentum." *The Journal of Investing* 16, no.1, 36–42.

Brush, J.S. and K.E. Boles. 1983. "The Predictive Power in Relative Strength & CAPM." The *Journal of Portfolio Management* 9(4), 20–23.

Carhart, M.M. 1997. "On Persistence in Mutual Fund Performance." *The Journal of Finance* 52, no.1, 57–82.

Carrillo-Gamboa, O.and R.F. Gunst.1992. "Measurement-Error-Model Collinearities." *Technometrics* 34, no.4, 454–464.

Ceria, S., K. Sivaramakrishnan, and R.A Stubbs. 2017. "Alpha Construction in a Consistent Investment Process." In J.B. Guerard, Jr., Ed. *Portfolio Construction, Measurement, and Efficiency: Essays in Honor of Jack Treynor.* New York: Springer.

Chan, L.K.C., Y. Hamao, and J. Lakonishok 1993. "Can Fundamentals Predict Japanese Stock Returns?" *Financial Analysts Journal* 49(4), 63–69.

Chan, L.K.C., Y. Hamao, and J. Lakonishok. 1991. "Fundamentals and Stock Returns in Japan." The *Journal of Finance* 46, no.5, 1739–1764.

Chandler, A.D., Jr. 1977. *The Visible Hand: The Managerial Revolution in American Business.* Cambridge, MA: Belknap Press, Chapter 13.

Chen, C. "Robust Regression and Outlier Detection with the ROBUSTREG Procedure." Paper 265-27. *Proceedings of the Twenty-Seventh* Annual SAS Users Group International Conference. Cary, NC: SAS Institute.

Chopra, V.K., and W.T. Ziemba. 1993. "The Effect of Errors in Means, Variances, and Covariances on Optimal Portfolio Choice." The *Journal of Portfolio Management* 19, no.2, 6–11.

Christopherson, J.A., D.R. Carino, and W. E. Ferson. 2009. *Portfolio Performance Measurement and Benchmarking.* New York: McGraw-Hill.

Cochran, W.G. 1934. "The distribution of quadratic forms in a normal system, with applications to the analysis of covariance." *Mathematical Proceedings of the Cambridge Philosophical Society* 30(2), 178–191.

Cochrane, J.H. and T.J. Moskowitz. 2017. *The Fama Portfolio: Selected Papers of Eugene F. Fama.* Chicago: The University of Chicago Press.

Conner, G. and R.A. Korajczyk. 1995. "The Arbitrage Pricing Theory and Multifactor Models of Asset Returns." In R. A. Jarrow, V. Maksimovic, and W.T. Ziemba, Editors, *Handbooks in Operations Research and Management Science: Finance* 9, 87–144. Amsterdam: Elsevier.

Cootner, P. 1964. *The Random Character of Stock Market Prices.* Cambridge: MIT Press.

Dewing, A.S. 1953. *The Financial Policy of Corporations*, 5th edition. New York: Ronald Press. Vol. 1. Chapters 1–4.

Dhrymes, P.J. and M. Kurz. 1967. "Investment, Dividends, and External Finance Behavior of Firms." In *Determinants of Investment Behavior*, edited by Robert Ferber. New York: Columbia University Press.

Dhrymes, P.J., I. Friend, and N.B. Gultekin. 1984. "A Critical Reexamination of the Empirical Evidence on the Arbitrage Pricing Theory." *The Journal of Finance* 39, No.2, 323–346.

Dhrymes, P.J., I. Friend, M.N. Gültekin, and N.B. Gültekin. 1985. "New Tests of the APT and Their Implications." *The Journal of Finance* 40, no.3, 659–674.

Dimson, E. 1988. *Stock Market Anomalies.* Cambridge: Cambridge University Press.

Dreman, D.N. 1979. *Contrarian Investment Strategy: the Psychology of Stock Market Success.* New York: Random House.

Dreman, D.N. 1998. *Contrarian Investment Strategies: The Next Generation: Beat the Market by Going Against the Crowd.* New York: Simon & Schuster.

Drtina, R.E. and J.A. Largay, III. 1985. "Pitfalls in Calculating Cash Flow from Operations." *The Accounting Review* 60, no.2, 314–326.

Efron, B. and Morris, C. 1973. "Stein's Estimation Rule and Its Competitors—An Empirical Bayes Approach." *Journal of the American Statistical Association.* 68(341), 117–130.

Elton, E. J., M.J. Gruber, and M.W. Padberg. 1979. "Simple Criteria for Optimal Portfolio Selection: The Multi-Index Case." In *Portfolio Theory, 25 Years After: Essays in Honor of Harry Markowitz,* edited by E.J. Elton and M.J. Gruber. Amsterdam: North-Holland.

Elton, E.J. and M.J. Gruber. 1970. "Homogeneous Groups and the Testing of Economic Hypotheses." *The Journal of Financial and Quantitative Analysis* v.4, no.5, 581–602.

Elton, E.J. and M.J. Gruber. 1973. "Estimating the Dependence Structure of Share Prices—Implications for Portfolio Selection." *The Journal of Finance* 28, no.5, 1203–1232.

Elton, E.J., M.J. Gruber, and M. Gültekin. 1981. "Expectations and Share Prices." *Management Science* 27, no.9, 975–987.

Elton, E.J., M.J. Gruber, and M. Padberg. 1978. "Simple Criteria for Optimal Portfolio Selection: Tracing out the Efficient Frontier." *The Journal of Finance* 33, no.1, 296–302.

Elton, E.J., M.J. Gruber, S.J. Brown, and W.N. Goetzmann. 2007. *Modern Portfolio Theory and Investment Analysis, Seventh Edition.* Hoboken, NJ: John Wiley & Sons.

Fabozzi, F.J., F. Gupta and H.M. Markowitz. 2002. "The Legacy of Modern Portfolio Theory." *The Journal of Investing* 11(3), 7–22.

Fama, E.F. 1970. "Efficient Capital Markets:A Review of Theory and Empirical Work." *The Journal of Finance* 25, 383–417.

Fama, E.F. 1976. *Foundations of Finance: Portfolio Decisions and Securities Prices.* New York: Basic Books.

Fama, E.F. 1991. "Efficient Capital Markets II." *The Journal of Finance* 46, 1575–1617.

Fama, E.F. and French, K.R. 2008. "Dissecting anomalies." The *Journal of Finance 63*, no.4, 1653–1678.

Fama, E.F. and J.D. MacBeth. 1973. "Risk, Return, and Equilibrium: Empirical Tests." *Journal of Political Economy* 81, no.3, 607–636.

Fama, E.F. and K. R. French. 1992. "The Cross-Section of Expected Stock Returns." *The Journal of Finance* 47, no.2, 427–465.

Fama, E.F. and K.R. French. 1995. "Size and Book-to-Market Factors in Earnings and Returns." *The Journal of Finance* 50, no.1, 131–155.

Fama, E.F. and M. Blume. 1966. "Filter Rules and Stock Market Trading Profits." *The Journal of Business* 39, 226–241.

Fama, E.F., L. Fisher, M.Jensen, and R. Roll. 1969. "The Adjustment of Stock Prices to New Information." *International Economic Review* 10, 1–21.

Farrell, J.L., Jr. 1974. "Analyzing Covariation of Returns to Determine Homogeneous Stock Groupings." *The Journal of Business* 47, no.2, 186–207.

Farrell, J.L., Jr. 1997. *Portfolio Management: Theory and Application.* 2nd ed. New York: McGraw-Hill.

Ferson, W.E. and C.R. Harvey. 1993. "Explaining the Predictability of Asset Returns." In *Research in Finance* Volume 11, edited by A.H. Chen. Bingley, UK: Emerald Group

Ferson, W.E. and R.W. Schadt. 1996. "Measuring Fund Strategy and Perormance in Changing Economic Conditions." *The Journal of Finance* 51, no.2, 425–461.

Finney, H.A., and H.E. Miller. 1958. *Principles of Accounting: Intermediate*, 5th edition. Englewood Cliffs, NJ: Prentice-Hall, Chapter 3.

Francis, J.C. and R. Ibbotson. 2002. Investments: A Global Perspective. Upper Saddle River, NJ: Prentice-Hall.

Graham, B. 1973. *The Intelligent Investor: a Book of Practical Counsel.* 4th rev. ed. New York: Harper & Row.

Graham, B. and D.L. Dodd. 1934. *Security Analysis.* New York: McGraw-Hill.

Graham, B., D.L. Dodd, and S. Cottle. 1934, 1962. *Security Analysis: Principles and Technique.* 1th and 4th Edition. New York: McGraw-Hill.

Grauer, R.R. and N.H. Hakansson. 1995. "Stein and CAPM estimators of the Means in Asset Allocation." *International Review of Financial Analysis* 4(1), 35–66.

Grinold, R.C. and R.N. Kahn. 2000. *Active Portfolio Management: a Quantitative Approach for Providing Superior Returns and Controlling Risk.* 2nd ed. New York: McGraw-Hill.

Gruber, M.J. 1996. "Another Puzzle: The Growth in Actively Managed Mutual Funds." *The Journal of Finance* 51, no.3, 783–810.

Grullon, G. and R. Michaely. 2001. "Dividends, Share Repurchases, and the Substitution Hypothesis." *Journal of Finance* 57, 1649–1684.

Guerard, J.B., Jr. 2010. "The Corporation as a Net Exporter of Funds: Additional Evidence." In R. J. Aronson, H. L. Parmet, and R. J. Thornton. Eds. *Variations in Economic Analysis: Essays in Honor of Eli Schwartz.* New York: Springer.

Guerard, J.B., Jr. 2012."Global Earnings Forecasting Efficiency." In *Research in Finance* Volume 28. Edited by J.W. Kensinger. Bingley, UK: Emerald.

Guerard, J.B., Jr. 2013. *Introduction to Financial Forecasting in Investment Analysis.* New York: Springer.

Guerard, J.B., Jr. 2016. "Investing in Global Markets: Big Data and Applications of Robust Regression." *Frontiers in Applied Mathematics and Statistics* 1, Article 14. 1:14. https://doi.org/10.3389/fams.2015.00014.

Guerard, J.B., Jr., and A.M. Mark. 2003. "The Optimization of Efficient portfolios: The Case for an R&D Quadratic Term." In *Research in Finance* Volume 20, A.H. Chen, editor. Bingley, UK: Emerald Group

Guerard, J.B., Jr., and A.M. Mark. 2018. "Earnings Forecasts and Revisions, Price Momentum, and Fundamental data: Further Explorations of Financial Anomalies." In C.F. Lee, J.C. Lee Ed. *Handbook of Financial Econometrics, Mathematics, Statistics, and Machine Learning.* Singapore: World Scientific Publishers, forthcoming.

Guerard, J.B., Jr., and B.K. Stone. 1992. "Composite Forecasting of Annual Earnings." In *Research in Finance* Volume 10. Edited by A.H. Chen. Greenwich, CT: JAI Press.

Guerard, J.B., Jr., and E. Schwartz. 2007. *Quantitative Corporate Finance.* New York: Springer.

Guerard, J.B., Jr., and H.M. Markowitz. 2019. "The Existence and Persistence of Financial Anomalies: What Have You Done for Me Lately?" *Financial Planning Review* 1, 2018; 1:e1022. https://doi.org/10.1002/cfp2.1022.

Guerard, J.B., Jr., A.S. Bean, and S. Andrews. 1987. "R and D Management and Corporate Financial Policy." *Management Science*, 33, 1419–142.

Guerard, J.B., Jr., G. Xu, and M. Gultekin. 2012. "Investing with momentum: the past, present, and future." *The Journal of Investing* 21, no.1, 68–80.

Guerard, J.B., Jr., Gultekin, M. and B.K. Stone. 1997. "The role of fundamental data and analysts' earnings breadth, forecasts, and revisions in the creation of efficient portfolios." In Research in Finance Volume 15. Edited by A.H. Chen.

Guerard, J.B., Jr., H.M. Markowitz, and G. Xu. 2014. "The Role of Effective Corporate Decisions in the Creation of Efficient Portfolios." *IBM Journal of Research and Development* 58(4),6.1–6.11.

Guerard, J.B., Jr., H.M. Markowitz, and G. Xu. 2015. "Earnings Forecasting in a Global Stock Selection Model and Efficient Portfolio Construction and Management." *International Journal of Forecasting* 31, no.2, 550–560.

Guerard, J.B., Jr., S.T. Rachev, and B.P. Shao. 2013. "Efficient Global Portfolios: Big Data and Investment Universes." *IBM Journal of Research and Development* 57(5), 11.1–11.11.

Gunst, R.F. and R.L. Mason. 1979. "Some Considerations in the Evaluation of Alternate Prediction Equations." *Technometrics,* 21, no.1, 55–63.

Gunst, R.F. and R.L. Mason. 1980. *Regression Analysis and its Application: a Data-Oriented Approach.* New York: Marcel Dekker.

Gunst, R.F., Webster, J.T., and R.L. Mason. 1976. "A comparison of least squares and latent root regression estimators." *Technometrics, 18*, no.1, 75–83.

Hansen, L.P. 1982. "Large Sample Properties of Generalized Method of Moments Estimators." *Econometrica* 50(4), 1029–1054.

Harvey, C.R. and Y. Liu. 2014. "Evaluating Trading Strategies." SSRN: https://papers.ssrn.com/sol3/papers.cfm?abstract_id=2474755

Harvey, C.R. and Y. Liu. 2014. "Lucky Factors." SSRN: https://papers.ssrn.com/sol3/papers.cfm?abstract_id=2528780

Haugen, R.A. 1999. *The Inefficient Stock Market: What Pays Off and Why*. Upper Saddle River, N.J.: Prentice Hall.

Haugen, R.A. 2001. *Modern Investment Theory*. 5th edition. Upper Saddle River, N.J.: Prentice Hall.

Haugen, R.A. and N.L. Baker. 1996. "Commonality in the Determinants of Expected stock returns." *Journal of Financial Economics* 41, no.3, 401–439.

Haugen, R.A. and N.L. Baker. 2010. "Case closed." In J.B. Guerard, Jr. (ed.), *Handbook of Portfolio Construction: Contemporary Applications of Markowitz Techniques*. New York: Springer.

Henriksson, R.D. and R.C. Merton. 1981. "On Market Timing and Investment Performance. II. Statistical Procedures for Evaluating Forecasting Skills." *The Journal of Business* 54, no.4, 513–533.

Holm, S. 1979. "A simple sequentially rejective multiple test procedure." *Scandinavian Journal of Statistics* 6, no.2, 65–70.

Horngren, C.T.1984. *Introduction to Financial Accounting*, Englewood Cliffs, NJ: Prentice-Hall, 2nd edition. Chapter 7.

Hunt, P., Williams, C. M., and G. Donaldson. 1961. *Basic Business Finance: Text and Cases*, rev. ed., Homewood, IL: Richard D. Irwin. Ch. 8.

Ibbotson, Roger G. 2018. *Stocks, Bonds, Bills and Inflation* (SBBI) Yearbook. Chicago: Duff & Phelps.

Jacobs, B. I. and K.N. Levy. 1988. "Disentangling Equity Return Regularities: New Insights and Investment Opportunities." *Financial Analysts Journal* 44, no.3, 18–43.

Jacobs, B. I. and K.N. Levy. 2017. *Equity Management: The Art and Science of Modern Quantitative Investing*. Second Edition. New York: McGraw-Hill.

Jacobs, B.I. and K.N. Levy. 2010. "Reflections on Portfolio Insurance, Portfolio Theory, and Market Simulation with Harry Markowitz." In J.B. Guerard, Jr., editor, *Handbook of Portfolio Construction: Contemporary Applications of Markowitz Techniques*. New York: Springer.

Jaffee, J. F. 1974. "Special Information and Insider Trading." *The Journal of Business* 47, 410–428.

Jagannathan, R. and R.A. Korajczyk. 1986. "Assessing the Market Timing Performance of Managed Portfolios." *The Journal of Business* 59, no.2, 217–235.

James, W. and C. Stein. 1961. "Estimation with quadratic loss." .Proceedings of the Fourth Berkeley Symposium on Mathematical Statistics and Probability, Volume 1,361–379. Berkeley, CA: University of California Press.

Jensen, M.C. 1968. "The Performance of Mutual Funds in the Period 1945–1964." The *Journal of Finance* 23, no.2, 389–416.

Jensen, M.C. 1972. "Optimal Utilization of Market Forecasts and the Evaluation of Investment Performance." In G. P. Szego and Karl Shell(eds.), *Mathematical Methods in Investment and Finance*. Amsterdam: Elsevier.

Jensen, M.C. and W.H. Meckling. 1976. "Theory of the Firm: Managerial Behavior, Agency Costs, and Ownership Structure." *Journal of Financial Economics* 3, no.4, 305–360.

Johnson, R.W. 1959. *Financial Management*. Boston: Allyn and Bacon, Chapter 5.

Jorion, Philippe. 1986. "Bayes-Stein Estimation for Portfolio Analysis." *The Journal of Financial and Quantitative Analysis* 21, no.3, 279–292.

King, B.F. 1966. "Market and Industry Factors in Stock Price Behavior." *The Journal of Business* 39, no.1, Part 2, 139–190.

Kosowski, R., A. Timmermann, R. Wermers, and H. White. 2006. "Can Mutual Fund 'Stars' Really pick Stocks? New Evidence from a Bootstrap Analysis." *The Journal of Finance* 61, no.6, 2551–2595.

Kuhn, M. and K. Johnson. 2013. *Applied Predictive Modeling*. New York: Springer.

Lakonishok, J., A. Shleifer, and R.W. Vishny. 1994. "Contrarian Investment, Extrapolation, and Risk." *The Journal of Finance* 49, no.5, 1541–1578.

Latane, H. A. 1959. "Criteria for choice among risky ventures." *Journal of Political Economy* 67, no.2, 144–155.

Latane, H.A., D. Tuttle, and C.P. Jones. 1975. *Security Analysis and Portfolio Management*. New York: The Ronald Press.

Ledoit, O. and M. Wolf. 2003. "Improved estimation of the covariance matrix of stock returns with an application to portfolio selection." *Journal of Empirical Finance* 10, no.5, 603–621.

Lee C.F. and S. Rahman. 1990. "Market Timing, Selectivity, and Mutual Fund Performance: An Empirical Investigation." *The Journal of Business* 63, no.2 261–278.

Lee, C.F., A.C. Lee, and N. Liu. 2010. "Alternative Model to Evaluate Selectivity and Timing Performance of Mutual Fund Managers: Theory and Evidence." In J.B. Guerard, Jr., editor, *Handbook of Portfolio Construction: Contemporary Applications of Markowitz Techniques*. New York: Springer.

Lintner, J. 1965. "Security Prices, Risk, and Maximal Gains from Diversification." *The Journal of Finance* 20(4), 587–615.

Lintner, J. 1965. "The Valuation of Risk Assets and the Selection of Risky Investments in Stock Portfolios and Capital Budgets." *The Review of Economics and Statistics* 47(1), 13–37.

Lo, A.W. 2002. "The Statistics of Sharpe Ratios." *Financial Analysts Journal* 58(4), 36–52.

Lo, A.W. 2017. *Adaptive Markets: Financial Evolution at the Speed of Thought*. Princeton, NJ: Princeton University Press.

Lo, A.W. and A.C. MacKinlay. 1990. "Data-Snooping Biases in Tests of Financial Asset Pricing Models." *The Review of Financial Studies* 3, no.3, 431–467.

Lorie, J. H. and V. Niederhoffer. 1968. "Predictive and Statistical Properties of Insider Trading." *The Journal of Law and Economics* 11, 35–53.

Maness, T.S. and J.T. Zietlow. 2005. *Short-Term Financial Management.* 3rd edition. Mason, Ohio: South-Western, Chapter 15.

Mansfield, E., W.B. Allen, N.A. Doherty, and K. Weigelt. 2002. *Managerial Economics: Theory, Applications, and Cases*, 5th edition. New York: W.W. Norton.

Mao, J.C.T. 1969. *Quantitative Analysis of Financial Decisions.* New York: Macmillan, Chapters 13 and 14.

Markowitz, H.M. 1952. "Portfolio Selection." The *Journal of Finance 7*, no.1, 77–91.

Markowitz, H.M. 1959. *Portfolio Selection: Efficient Diversification of Investments.* Cowles Foundation for Research in Economics, Monograph No.16. New Haven, CT: Yale University Press.

Markowitz, H.M. 1976. "Investment for the Long Run: New Evidence for an Old Rule." *The Journal of Finance* 31, no.5, 1273–1286.

Markowitz, H.M. 2000. *Mean-Variance Analysis in Portfolio Choice and Capital Markets.* New Hope, PA: Frank J. Fabozzi Associates.

Markowitz, H.M. and G.L. Xu 1994. "Data Mining Corrections." The *Journal of Portfolio Management, 21*, no.1, 60–69.

Markowitz, H.M. and K. Blay. 2013. *Risk-Return Analysis: The Theory and Practice of Rational Investing.* New York, McGraw-Hill.

Maronna, R.A., R.D. Martin, and V.J. Yohai. 2006. *Robust Statistics: Theory and Methods.* Chichester, UK: John Wiley.

Maronna, R.A., R. D. Martin, V.J. Yohai, and M. Salibian-Barrera. 2019. *Robust Statistics: Theory and Methods (with R).* Second Edition. New York: Wiley.

Mason, R.L. and R.F. Gunst. 1985. "Outlier-Induced Collinearities." *Technometrics* 27, no.4, 401–407.

McInnes, J.M. and Carleton, W.J.. 1982. "Theory, Models and Implementation in Financial Management." *Management Science* 28, no.9, 957–978.

Menchero, J. and Z. Nagy. 2017. "Performance of Earnings Yield and Momentum Factors in US and International Equity Markets." In J.B. Guerard, Jr., Ed. *Portfolio Construction, Measurement, and Efficiency: Essays in Honor of Jack Treynor.* New York: Springer.

Menchero, J., A. Morozov, and P. Shepard. 2010. "Global Equity Risk Modeling." In J.B. Guerard, Jr., editor, *Handbook of Portfolio Construction: Contemporary Applications of Markowitz Techniques.* New York: Springer.

Miller, M.H. 1977. "Debt and Taxes." *The Journal of Finance* 32, no.2, 261–275.

Miller, M.H. and D. Orr. 1966. "A Model of the Demand for Money by Firms." *The Quarterly Journal of Economics* 80, no.3, 413–435.

Miller, W., G. Xu, and J.B. Guerard, Jr. 2014. "Portfolio Construction and Management in the Barra Aegis System: A Case Study using the USER Data." The *Journal of Investing* 23, no.4, 111–120.

Mossin, J. 1966. "Equilibrium in a Capital Asset Market." *Econometrica* 34, no.4, 768– 783.

Mossin, J. 1973. *Theory of Financial Markets.* Englewood Cliffs, NJ: Prentice-Hall.

Nerlove, M. 1968. "Factors Affecting Differences Among Rates of Return on Investments in Individual Common Stocks." *The Review of Economics and Statistics* 50, no.3, 312–331.

Newey,W.K. and K.D. West, 1987. "A Simple, Positive Semi-Definite, Heteroskedasticity and Autocorrelation Consistent Covariance Matrix." *Econometrica*, 55, no.3, 703–708.

OCED. 2018. "Household Financial Assets." https://data.oecd.org/hha/household-financial-assets.htm

Palepu, K.G., P.M. Healy, and V.L. Bernard. 2000. *Business Analysis & Valuation: Using Financial Statements: Text & Cases.* 2nd edition. Cincinnati, OH: South-Western College Publishing, Chapters 4, 5, 8.

Penman, S.H. 2001. *Financial Statement Analysis and Security Valuation.* Boston: McGraw-Hill/Irwin, Chapter 7.

Pfleiderer, P. and Bhattacharya, S.. 1983. "A Note on Performance Evaluation." *Technical Report 714*, Stanford, Calif.: Stanford University, Graduate School of Business.

Pogue, G.A. and R.N. Bussard. 1972. "A Linear Programming Model for Short Term Financial Planning Under Uncertainty." *Sloan Management Review* 13, no.3,69–98.

Ramnath, S., S. Rock, and P. Shane. 2008. "The Financial Analyst Forecasting Literature: A Taxonomy with Suggestions for Further Research." *International Journal of Forecasting* 24, no.1, 34–75.

Rao, C.R. 1959. "Some problems involving linear hypotheses in multivariate analysis." *Biometrika*, 46, no.1/2, 49–58.

Rao, C.R. 1973. *Linear Statistical Inference and Its Applications.* 2nd edition. New York: John Wiley.

Reinganum, M.R. 1981. "The Arbitrage Pricing Theory: Some Empirical Results." The *Journal of Finance* 36, no.2, 313–321.

Rey, T., A. Kordon, and C. Wells. 2012. *Applied Data Mining for Forecasting Using SAS.* Cary, NC: SAS Institute.

Richardson, S.A., R.G. Sloan, M.T. Soliman, and I. Tuna. 2001. "Information in Accruals about the Quality of Earnings." University of Michigan Business School Working Paper.

Roll, R. 1969. "Bias in Fitting the Sharpe Model to Time Series Data." *The Journal of Financial and Quantitative Analysis* 4, no.3, 271–289.

Roll, R. 1977. "A Critique of the Asset Pricing Theory's Tests Part I: On Past and Potential Testability of the Theory." *Journal of Financial Economics* 4, no.2, 129–176.

Roll, R. and Ross, S.A. 1980. "An Empirical Investigation of the Arbitrage Pricing Theory." *The Journal of Finance* 35, no.5, 1073–1103.

Rosenberg, B. 1974. "Extra-Market Components of Covariance in Security Returns." *The Journal of Financial and Quantitative Analysis* 9, no.2, 263–274.

Rosenberg, B. and A. Rudd. 1977. "The Yield/Beta/Residual Risk Tradeoff." Working paper no. 66. Research Program in Finance. Institute of Business and Economic Research. University of California, Berkeley.

Rosenberg, B. and V. Marathe. 1979. "Tests of Capital Asset Pricing Hypotheses." In *Research in Finance* Volume 1, edited by H.M. Levy. Greenwich, CT: JAI Press.

Rosenberg, B. and W. McKibben. 1973. "The Prediction of Systematic and Specific Risk in Common Stocks." *The Journal of Financial and Quantitative Analysis* 8, no.2, 317–333.

Rosenberg, B., M. Hoaglet, V. Marathe and W. McKibben. 1975. "Components of Covariance in Security Returns." Working paper no. 13, Research Program in Finance. Institute of Business and Economic Research. University of California. Berkeley.

Ross, S.A. 1976. "The Arbitrage Theory of Capital Asset Pricing." *Journal of Economic Theory* 13, 341–360.

Rousseeuw, P.J. 1984. "Least Median of Squares Regression." *Journal of the American Statistical Association* 79, no.388, 871–880.

Rousseeuw, P.J. and A.M. Leroy. 1987. *Robust Regression and Outlier Detection*. New York: John Wiley .

Rousseeuw, P.J. and K. Van Driessen. 1999. "A Fast Algorithm for the Minimum Covariance Determinant Estimator." *Technometrics* 41, no.3, 212–223.

Rousseeuw, P.J. and K. Van Driessen. 2000. "An Algorithm for Positive-Breakdown Regression Based on Concentration Steps." *Data Analysis: Scientific Modeling and Practical Application*, ed. W. Gaul, O. Opitz, and M. Schader, New York: Springer, 335–346.

Rousseeuw, P.J. and V. Yohai. 1984. "Robust Regression by Means of S-Estimators." In *Robust and Nonlinear Time Series Analysis*, ed. J. Franke, W. Härdle, and R. D. Martin, Lecture Notes in Statistics, 26, New York: Springer-Verlag, 256–272.

Rubinstein, M.E. 1973. "A Mean-Variance Synthesis of Corporate Financial Theory." *The Journal of Finance* 28, no.1,167–181.

Rudd, A. and B. Rosenberg. 1979. "Realistic Portfolio Optimization." In *Portfolio Theory, 25 Years After*. Edited by E.J. Elton and M.J. Gruber. Amsterdam: North-Holland.

Rudd, A. and B. Rosenberg. 1980. "The 'Market Model' in Investment Management." *The Journal of Finance* 35, no.2, 597–607.

Rudd, A. and H.K. Clasing, Jr. 1982. *Modern Portfolio Theory: The Principles of Investment Management.* Homewood, Illinois: Dow Jones-Irwin.

Ruppert, D. 1992. "Computing S Estimators for Regression and Multivariate Location/Dispersion." *Journal of Computational and Graphical Statistics* 1, no.3, 253–270.

Saxena, A. and R.A. Stubbs 2015. "Augmented Risk Models to Mitigate Factor Alignment Problems." *Journal of Investment Management* 13, no.3, 57–79.

Saxena, A. and R.A. Stubbs. 2012. "An Empirical Case Study of Factor Alignment Problems using the USER Model." *The Journal of Investing* 21, no.1, 25–43.

Schwartz, E. 1959. "Theory of the Capital Structure of the Firm." *The Journal of Finance* 14, no.1,18–39.

Schwartz, E. 1962. *Corporation Finance*. New York: St. Martin's Press, Chapter 3.

Schwartz, E. and J.R. Aronson. 1966. "The Corporate Sector:A Net Exporter of Funds." *Southern Economic Journal* 33, no.2, 252–257.

Scott, J.H. 1976. "A Theory of Capital Structure." *The Bell Journal of Economics* 7, no.1, 33–54.

Sharpe, W.F. 1963. "A Simplified Model for Portfolio Analysis." *Management Science* 9(2), 277–293.

Sharpe, W.F. 1964. "Capital Asset Prices: A Theory of Market Equilibrium under Conditions of Risk." *The Journal of Finance* 19, no.3, 425–442.

Sharpe, W.F. 1966. "Mutual Fund Performance." *The Journal of Business* 39, no.1, Part 2, 119–138.

Sharpe, W.F. 1970. Portfolio Theory and Capital Markets. New York: McGraw-Hill.

Sharpe, W.F. 1971. "A Linear Programming Approximation for the General Portfolio Analysis Problem." *The Journal of Financial and Quantitative Analysis 6(5)*, 1263–1275

Sharpe, W.F. 1971. "Mean-Absolute-Deviation Characteristic Lines for Securities and Portfolios." *Management Science* 18(2), B1–B13.

Sivaramakrishnan, K. and R.A. Stubbs. 2013. "Improving the Investment Process with a Custom Risk Model: A Case Study with the GLER Model." *The Journal of Investing* 22, no.4, 129–147.

Sloan, R.G. 1996. "Do Stock Prices Reflect Information in Accruals and Cash Flows about Future Earnings?" *The Accounting Review* 71, no.3, 289–315.

Solomon, E. 1963. *The Theory of Financial Management.* New York: Columbia University Press.

Stein, C. 1956. "Inadmissibility of the usual estimator for the mean of a multivariate normal distribution." Proceedings of the Third Berkeley Symposium on Mathematical Statistics and Probability, Volume 1, 197–206. Berkeley, CA: University of California Press.

Stone, B.K. 1970. *Risk, Return, and Equilibrium: A General Single-Period Theory of Asset Selection and Capital-Market Equilibrium*. Cambridge, MA: MIT Press.

Stone, B.K. 1973. "A Linear Programming Formulation of the General Portfolio Selection Problem." *The Journal of Financial and Quantitative Analysis* 8, no.4, 621–636.

Stone, B.K. 1974. "Systematic Interest-Rate Risk in a Two-Index Model of Returns." *The Journal of Financial and Quantitative Analysis* 9, 709–721.

Stone, B.K. and J.B. Guerard, Jr. 2010. "Methodologies for Isolating and Assessing Potential Portfolio Performance of Stock Return Forecast Models with an Illustration." In J.B. Guerard, Jr., editor, *The Handbook of Portfolio Construction: Contemporary Applications of Markowitz Techniques*. New York: Springer.

Tobin, J. 1958. "Liquidity preference as behavior towards risk." The *Review of Economic Studies*." 25(2):65–86.

Torry, Harriet. 2018. "Americans' Wealth Surpasses $100 Trillion." The Wall Street Journal. https://www.wsj.com/articles/u-s-net-worth-surpasses-100-trillion-1528387386

Treynor, J.L. 1965. "How to Rate Management of Investment Funds." *Harvard Business Review* 43, 63–75.

Treynor, J.L. 1993. "In Defense of the CAPM." *Financial Analysts Journal* 51, 93–100.

Treynor, J.L. 1999. "Toward A Theory of Market Value for Risky Assets." In *Asset Pricing and Portfolio Performance*, ed. Robert Korajczyk. Risk Books.

Treynor, J.L. and K.K. Mazuy. 1966. "Can Mutual Funds Outguess the Market." *Harvard Business Review*, 44, no.4, 131–136.

Treynor, J.L., W.W. Priest, Jr., L. Fisher, and C.A. Higgins. 1968. "Using Portfolio Composition to Estimate Risk." *Financial Analysts Journal* 24, no.5, 93–100.

United States Equity. 1998. Version 3 (E3), BARRA. *Risk Model Handbook.* http://www.alacra.com/alacra/help/barra_handbook_US.pdf

Van Horne, J.C. 2002. *Financial Management & Policy*. 12th edition. Upper Saddle River, New Jersey: Prentice-Hall, Chapter 1.

Van Horne, J.C. and J.M. Wachowicz. 2001. *Fundamentals of Financial Management*. Eleventh Edition. Upper Saddle River, New Jersey: Prentice-Hall, Chapter 6.

Vasicek, O.A. 1973. "A Note on Using Cross-Sectional Information in Bayesian Estimation of Security Betas." *The Journal of Finance* 28, no.5, 1233–1239.

Wei, K. D., R. Wermers, and T. Yao. 2015. "Uncommon Value: The Characteristics and Investment Performance of Contrarian Funds." *Management Science* 61, no.10, 2394–2414.

Wermers, R. 1999. "Mutual Fund Herding and the Impact on Stock Prices." *The Journal of Finance* 54, no.2, 581–622.

Weston, J.F. and T.E. Copeland. 1986. *Managerial Finance,* 8th edition. Chicago: Dryden Press. Chapters 2, 11, 26.

White, G.I., Sondhi, A.C., and D. Fried. 1998. *The Analysis and Use of Financial Statements*. 2nd edition. New York: John Wiley . Chapters 2, 4.

White, H. 1984. *Asymptotic Theory for Econometricians*, Orlando: Academic Press.

Williams, J.B. 1938. *The Theory of Investment Value*. Cambridge, MA: Harvard University Press.

Xu, G. 2015. "The Risk Profiles of 401(k) Accounts." *The Journal of Retirement* 2, no.3, 67–77.

Yohai V.J. 1987. "High Breakdown-Point and High Efficiency Robust Estimates for Regression." *The Annals of Statistics* 15, no.2, 642–656.

Yohai V.J., Stahel, W.A. and Zamar, R.H. 1991. "A Procedure for Robust Estimation and Inference in Linear Regression." In Stahel, W. and Weisberg, S., eds., *Directions in Robust Statistics and Diagnostics, Part II*, New York: Springer-Verlag.

Yohai, V.J. and Zamar, R.H. 1997. "Optimal Locally Robust M-estimates of Regression." *Journal of Statistical Planning and Inference* 64, no.2, 309–323.

Ziemba, W.T. 2010. "Ideas in Asset and Asset-Liability Management in the Tradition of H.M. Markowitz." In J.B. Guerard (ed.), *Handbook of Portfolio Construction: contemporary Applications of Markowitz Techniques*. New York: Springer.

Ziemba, W.T. and S. Schwartz. 1993. *Invest Japan*. Chicago: Probus.

www.ingramcontent.com/pod-product-compliance
Lightning Source LLC
Chambersburg PA
CBHW061324190326
41458CB00011B/3893

* 9 7 8 1 6 4 2 9 5 1 9 3 6 *